ECE/TIM/DP/22

United Nations Economic Commission for Europe
Food and Agriculture Organization
of the United Nations

Timber Section,
Geneva, Switzerland

Geneva Timber and Forest Discussion Papers

STRUCTURAL, COMPOSITIONAL AND FUNCTIONAL ASPECTS OF FOREST BIODIVERSITY IN EUROPE

by

Dr Janna Puumalainen

Environment and GEOinformation Unit, JRC - European Commission

UNITED NATIONS

New York and Geneva, 2001

NOTE

The designation employed and the presentation of material in this publication do not imply the expression of any opinion whatsoever on the part of the secretariat of the United /nations concerning the legal status of any country, territory, city or area, or of its authorities, or concerning the delimitation of its frontiers or boundaries.

LEGAL NOTICE

Neither the European Commission nor any person acting on behalf of the Commission is responsible for the use which might be made of the following information.

This study also has the following European Commission reference code:
EUR 19904 EN

ECE/TIM/DP/22

UNITED NATIONS PUBLICATION

Sales No. E.01.II.E.21

ISBN 92-1-116788-4

ISSN 1020-7228

UNECE/FAO TIMBER AND FOREST DISCUSSION PAPERS

The objective of the Discussion Papers is to make available to a wider audience work carried out, usually by national experts, in the course of ECE/FAO activities. They do not represent the final official output of the activity, but rather a contribution which because of its subject matter, or quality, or for other reasons, deserves to be disseminated more widely than the restricted official circles from whose work it emerged, or which is not suitable (e.g. because of technical content, narrow focus, specialized audience) for distribution in the UNECE/FAO Timber and Forest Study Paper series.

In all cases, the author(s) of the discussion paper are identified, and the paper is solely their responsibility. The ECE Timber Committee, the FAO European Forestry Commission, the governments of the authors' country and the FAO/ECE secretariat, are neither responsible for the opinions expressed, nor the facts presented, nor the conclusions and recommendations in the discussion paper.

In the interests of economy, Discussion Papers are issued in the original language only. They are available on request from the secretariat. They are distributed automatically to nominated forestry libraries and information centres in member countries. It is the intention to include this discussion paper on the Timber Committee website at: http//www.unece.org/trade/timber. Those interested in receiving these Discussion Papers on the continuing basis should contact the secretariat.

Another objective of the Discussion Papers is to stimulate dialogue and contacts among specialists. Comments or questions should be sent to the secretariat, who will transmit them to the authors.

Table of Contents

Foreword and Acknowledgements

The forestry world is being flooded by biodiversity information related to different disciplines, geographical areas, scales of observation, organisms and, not the least, to different interests of various stake holders. In practice, international assessments and monitoring still pose some difficulties, which are often related to unavailability and incompatibility of the data, to different standards and definitions and to restricted assessment schemes, either geographically or thematically. Also, the lack of cooperation between different initiatives may hinder holistic assessments. These were some of the accounted problems, which initiated the preparation of the current paper.

The paper has evolved as a joint effort of the UNECE secretariat in Geneve and the Joint Research Centre of the European Commission in Ispra. In practice, the paper has been compiled within the EUROLANDSCAPE Project at the Environment and GEOinformation Unit of the Joint Research Centre in close collaboration with the TBFRA-2000 Team of the UNECE/FAO secretariat. The team has also made the data available for the study. Especially the very important contribution of Pamela Kennedy and Sten Folving (EGEO/JRC) and Alexander V. Korotkov (UNECE/FAO secretariat) has been crucial during the whole preparation of the document.

The draft manuscript has been reviewed and valuably commented by Tor-Björn Larsson (Swedish Environmental Protection Agency), Jari Parviainen (Finnish Forest Research Institute), Hans Pretzsch (Forestry Department of the Technical University of Munich, Germany) and Yves Zanatta (Eurostat, Luxembourg). Their remarks and suggestions have greatly improved the quality of the paper, and I really appreciate the considerable effort and input I received.

Janna Puumalainen
Environment and GEOinformation Unit
JRC - European Commission
E-mail: janna.puumalainen@jrc.it

Countries and Country Groupings

The original TBFRA-2000 assessment covered fifty-five countries in Europe, CIS, North America, Australia, Japan and New Zealand. This report concentrates solely on Europe. Two main groupings of countries are distinguished in the analysis: whole Europe and the fifteen countries of the European Union (EU-15). The term "Pan-European" is used solely in connection with the Pan-European process and the criteria and indicators developed within this process (i.e. Ministerial Conference on the Protection of Forests in Europe, MCPFE).

All the European countries (Europe) included in the analysis are (in alphabetical order):

Albania, Austria, Belgium, Bosnia and Herzegovina, Bulgaria, Croatia, Cyprus, Czech Republic, Denmark, Estonia, Finland, France, Germany, Greece, Hungary, Iceland, Ireland, Israel, Italy, Latvia, Liechtenstein, Lithuania, Luxembourg, Malta, the Netherlands, Norway, Poland, Portugal, Romania, Slovakia, Slovenia, Spain, Sweden, Switzerland, the former Yugoslav Republic (FYR) of Macedonia, Turkey, United Kingdom and Yugoslavia.

The EU-15 countries are (in alphabetical order):

Austria, Belgium, Denmark, Finland, France, Germany, Greece, Ireland, Italy, Luxembourg, the Netherlands, Portugal, Spain, Sweden and the United Kingdom.

Abbreviations and Definitions

Abbreviations

CIS	Commonwealth of Independent States
CBD	Convention on Biological Diversity
EC	European Commission
ECE (UNECE)	United Nations Economic Commission for Europe
EEA	European Environment Agency
EFI	European Forest Institute
EFICS	European Forest Information and Communication System
EU	European Union
FAO	Food and Agriculture Organization of the United Nations
FAWS	Forest available for wood supply
FNAWS	Forest not available for wood supply
FOWL	Forest and other wooded land
FRA	Forest Resources Assessment(s)
ICP	Forest International Co-operative Programme on the Assessment and Monitoring of Air Pollution Effects on Forests
IUCN	International Union for the Conservation of Nature and Natural Resources
IUFRO	International Union of Forest Research Organizations
JRC	Joint Research Centre of the European Commission
MCPFE	Ministerial Conference on the Protection of Forests in Europe
OWL	Other wooded land
SAI/JRC	Space Application Institute of the Joint Research Centre
TBFRA-2000	Temperate and boreal forest resources assessment (see UN 2000 in references)
UNCED	United Nations Conference on Environment and Development
UK	United Kingdom of Great Britain and Northern Ireland
WWF	World Wide Fund for Nature
WCMC	World Conservation Monitoring Centre

0	nil or less than half a unit
cm	centimetre
m	metre
m^3	cubic metre, solid volume
m^3 o.b.	cubic metre, overbark
m^3 u.b.	cubic metre, underbark
ha	hectare
km^2	square kilometer

Definitions

Throughout the paper the terms and definitions applied in the UNECE/FAO Temperate and Boreal Forest Resources Assessment 2000 are used (UN 2000). Only in some cases where seen necessary the definitions are provided in the text of this paper. The complete list of terms and definitions is located in the Appendix I of the TBFRA-2000 Main Report (http://www.unece.org/trade/timber/fra/screen/append1.pdf).

List of Tables

List of Figures

Summary

International reporting, large-scale assessments and long-term monitoring of forest biodiversity gain a lot of attention, but pose still difficulties to overcome. Some assessments concentrate on a few selected indicators, some on only a fraction of the geographical area and some use data, which are not directly comparable due to different definitions and standards used in national assessments. Joint analysis of the data of different initiatives has not always been possible, and political processes on the selection of criteria and indicators may have received attention instead of the plain assessment. Due to these reasons there is no comprehensive overview of the forest biodiversity and variety of the forests in Europe. Thus, this paper attempts to make justice to forest biodiversity as a multidimensional issue and, furthermore, to the variety of European forests. The paper aims to provide a comprehensive overview of the European forest biodiversity. Subsequent to the assessment of the current state the paper aims to analyse the difficulties and deficiencies related to such assessments, and further to outline improvements and alternatives for the future. A pragmatic approach has been taken so that despite of missing universal definition of biodiversity, non-existing "final" list of biodiversity indicators and lacking "perfect" data, an assessment of European forest biodiversity has been made.

Resolution, data and methods to describe biodiversity

The area of European forests is not decreasing. Thus, the loss of biodiversity appears in the qualitative aspect of the forests, in changes of the forests structure, composition and functions. Several indicators are needed for a comprehensive biodiversity assessment – and no final list of indicators currently exists. Both the Pan-European indicators on sustainable forest management (e.g. MCPFE 2000) and especially the framework of structural, compositional and functional key factors of European biodiversity (Larsson et al. 2001) are applied in this work. The relationships between the two approaches are evaluated, and the structural, compositional and functional key factors are considered as complementary to the Pan-European indicators in that they provide an extensive framework for the analysis of forest biodiversity.

Among the structural factors emphasis is given especially on the natural and protected forests and on structurally diverse forests (i.e. mixed species and uneven-aged forests) in Europe. Compositional key factors include a validation of species numbers from different data sources, estimation of the importance of forests as a habitat for species and threatened species and an analysis of species richness per unit area. The functional aspect of forest biodiversity was found to be the most vague and several difficulties related to the concrete appraisal were identified. Thus, the presentation of this component is the most limited one in this paper.

The analysis is carried out at country level and comprises geographically the European countries covered by the Temporal and Boreal Forest Research Assessment (TBFRA-2000). The latest results from TBFRA-2000 (UN 2000) have been available for the analysis, and they have been complemented or verified with additional sources when appropriate (e.g. protected areas and species' numbers). The availability and location of the TBFRA-2000 data with respect to biodiversity indicators are also listed.

Suitability of the TBFRA-2000 data for biodiversity assessments

For an evaluation of biodiversity the ecosystem composition, structure and processes need to be monitored at different scales ranging from continent and stand level to a scale of an organism, and from centuries and years down to months and days. The dimension and need for these different scales is acknowledged, while the current study concentrates on medium and large-scale elements relevant at regional or continental assessments. The TBFRA-2000 data serves this purpose well – despite of some shortcomings with respect to the assessment of "new" variables related to the quality of forests. Especially the application of uniform definitions and standards should be credited.

The TBFRA-2000 directly provides data on eight of the twenty quantitative Pan-European indicators on sustainable forest management and five further indicators are covered partly or through data modifications. Furthermore, the TBFRA-2000 provides information covering almost all of the structural, compositional and functional key factors – at least to some extent. Care is needed especially when interpreting the data on species numbers and protected areas. To some degree, different interpretations may affect the data on plantations, mixed forests and uneven-aged forests. Generally, the change estimation is problematic due to the severe methodological difficulties related to monitoring.

Natural and protected forests

There are hardly any true natural forests remaining in Europe. Only about 4 % of the forest area has been classified as undisturbed by man in the TBFRA-2000 – mainly in countries located in the northern or eastern Europe. Most of these remaining natural forests are under some kind of protection scheme. Generally, western, northern and eastern European countries have somewhat different approach to the protection of forests. The non-EU countries especially in Eastern Europe have a larger percentage of forest area under the strict protection categories than the EU-countries. Only Finland, (Sweden), Italy and Portugal form an exception in this pattern.

The situation of forest protection in Europe is in no way simple and clear. There are nearly 90 different categories of protected forests ranging from national parks to aesthetic forests, and the forest protection policy and implementation differ widely between countries. The average percentage of protected areas of forest and other wooded land is 21.6% (equal to 32 million ha) according to the TBFRA-2000 and 7.7% (equals to 11 million ha) according to the COST Action E4 on Forest Reserves Research Network (Parviainen et al. 2000). In the percentage of the strict protected areas there is also a difference: 3.6% of the forests and other wooded land has been classified in the strictly protected IUCN I-II categories in the TBFRA-2000 and 0.9% in the COST Action E4 (based on the data from 21 countries). Different definitions of forest protection explain part of the differences in the estimated figures, but not the whole extent of the variation. Thus, detailed and meaningful comparisons of the protected forests within Europe require further clarification and analysis.

Structurally diverse forests

Mixed species and mixed age classes imply structural diversity within a forest. The current percentage of mixed species forests amounts to 14% of forest area in Europe, and is somewhat lower if the EU-countries are investigated alone (8.5%) (TBFRA-2000). Coppices cover about 21 million ha or 16% of forests available for wood supply, mainly in Southern and Southeastern Europe. Coppice structures are contrasted with high forests. Uneven-aged forests comprise 16% of the high forests. The percentage of uneven-aged mixed forests is moderate – on average less than 4% of the high forest available for wood supply.

Number of species in the TBFRA-2000 and other international assessments

Composition as an element of biodiversity relates most commonly to the number of species. In the TBFRA-2000 assessment information has been collected not only on the total number of different species, but also on the number of forest occurring species, and separate data have been collected on the number of different tree species. These data are complementary to most other international processes, which mainly collect information on the species number, independent of their habitat.

Even if the number of species seems to be a simple indicator at the first glance, the reported numbers are not in all cases consistent and therefore interpretations should be done carefully. Reasons for the varying figures are different interpretations in terms of the geographical area, taxonomical groupings, endangered categories, and all kinds of misinterpretations. The number of faunal species seems to vary providing if only regularly residing species or the total recorded number has been provided. An extensive attempt has been made to verify, compare and complement the figures reported in the TBFRA-2000 by using figures reported in other international processes (especially WCMC 1994 and EC 1995), recent reports on national biodiversity and other data sources.

Number of species depending on forests in Europe

The European contribution to the total number of species in the world is relative moderate. Only some 2-6% of the world's species are present in Europe, varying according to the species group (EEA 1999). The proportion of species occurring only in Europe, however, is considerable (1/3 – ¾) for several species groups. How important forests are as a habitat for different species, i.e. how large a proportion of other vascular plants, ferns, mosses, lichens, mammals, birds, other vertebrates and butterflies and moths reside in forests, varies a lot between the countries and depends on the organism in question. Generally, the role of forests as species habitat seems to increase towards the East in Europe, independent of the species group studies.

In most countries 10-30% of vascular plant species and at least 50% of the fern and mammal species are forest-occurring. Generally birds seem to be less dependent on forests as a habitat than for instance mammals. The number of forest-occurring bird species increases towards the North and East in Europe. Forests seem to be a relatively "good" habitat in the sense that for most taxonomic groups (especially vascular plants and ferns) the species tend to be proportionally less endangered in forests than in other habitats. For instance in France and Germany about 16-19% of vascular plants are found in forests, but only 2-4% of the endangered vascular plants are forest-occurring. Forest ecosystems and other habitats seem to contribute equally to the number of endangered mammal species.

Species richness and geographical location

Globally Europe does not belong to "hotspots" of species richness, especially if the Mediterranean basin is excluded. Generally species diversity decreases with increasing latitude, even at the European scale: The estimated number of vascular plants in a country is over 9,000 in Turkey, around 5,500 in Italy, 4,500 in France, 3,200 in Germany and around 1,200 in Finland. The estimated number of mammals in a country ranges from over 130 species in Turkey, around 100 species in Italy and France, 90 species in Germany to a bit over 60 species in Finland.

If the total number of forest-occurring species, be it trees, other vascular plants, birds or mammals, is related to the unit area, i.e. is divided by the area of forests and other wooded land in the country, the small European countries are the most "species-rich" countries, independent of the factor in question. This is apparently the wrong scale to locate European species diversity hotspots. It is more important to locate the valuable areas within each country, forest type and bio-geographical region – rather than to do evaluations at a country level.

Holistic assessment of forest biodiversity

Generally it becomes clear that in different European countries various aspects of structural, compositional and functional variety are accentuated in different ways. This is due to the rich variety of natural conditions and ecological factors, but considerably also due to current and past anthropogenic changes. This variety even at such a low level of resolution like the national level emphasizes the fact that one or only a few indicators can rarely suffice and describe the true variety in the forests in Europe and that the attempts to assign one comparable value for the forest biodiversity are very difficult. It is also obvious that national level reporting is not an early-warning system and should be complemented by more detailed assessments.

Holistic approach of assessing environmental problems includes no only the monitoring at several scales and by using a variety of different parameters but also an integrated analysis of ecological, social and economic phenomena. The history of the forests, landscape, people and their interactions is rich and complex in Europe. When the dependencies between the socio-economical development and biodiversity are studied, one deals with the fundamental – and not with the proximate – causes affecting the biodiversity. Only an integrated approach in understanding the complex interdependencies between the socio-economic development and natural component would allow a proper understanding and prognosis of the processes underlying the depletion of biodiversity.

1. Introduction

What is actually the status of European forest biodiversity? Something like "All in all, there is little diversity in European forests today, with just a few species dominating..." (EEA 1995)? Or "The richness and diversity of rural landscapes in Europe is a distinctive feature of the continent. There is probably nowhere else where the signs of human interaction with nature in landscape are so varied, contrasting and localised." (Dobris Assessment, EC 1995). Or something else? This paper aims to bring some light to the status, description and monitoring of the biodiversity within European forests. Since forest biodiversity is a multidimensional issue and Europe comprises different bio-geographical regions, varying landscape histories and numerous different forest types, a broad analysis is required to adequately assess this variety.

International initiatives on forest biodiversity

Since the United Nations Conference on Environment and Development (UNCED 1992) in Rio de Janeiro the biodiversity issues have been visible in the international agenda. Forests are the major contributor to the terrestrial biodiversity in the world, and even in Europe where the human impact has a long history, forests are still ecologically far less disturbed than the areas of other land uses. Some common changes, which have occurred in forests during the last decades include loss of old forests, simplification of forest structure, decreasing size of forest patches, increasing isolation of patches, disruption of natural fire regimes, increased road building (e.g. Noss 1999) and introduction of non-native or genetically modified and selected species as well as large plantations of conifers and Eucalyptus (Nowicki et al. 1998).

In the context of the Rio Conference most of the European countries have signed the "Convention of Biological Diversity", and there are several European or European Union activities related to the implementation of the convention into forestry and forest biodiversity. The main ministerial level processes are "The Pan-European process" ("Helsinki-process") on the protection of forests in Europe and the "Environment for Europe". In the former process the Resolution H2 "General guidelines for the conservation of biodiversity of European forest" was adopted at the Helsinki Conference, and the latter has defined the "Pan-European Biological and Landscape Diversity Strategy".

Further European level activities include the "Biodiversity Strategy" and "Communication on a forestry strategy for the European Union (EC 1998)" developed by the European Commission. In 2001 the European Commission adopted biodiversity action plans to integrate the protection of biodiversity in the areas of conservation of natural resources, agriculture, fisheries, and development and economic co-operation (EC 2001). Forest biodiversity issues are also to some extent taken into account in the implementation of the EU Habitats directive and Natura 2000. The European Environmental Agency (EEA) is monitoring the state of the European environment, including some aspects of forest biodiversity. Numerous regional and national initiatives are also in operation or planned for supporting sustainable development of the forests and forest biodiversity[1].

International reporting on biodiversity

International initiatives aim to maintain and improve the state of biodiversity. Hence, they need information for the assessment of the current state, selection of appropriate measures and monitoring of progress in the future. Without quantitative measures it is difficult to set concrete targets or formulate management regimes. Further, quantitative measures enable the monitoring and the assessment of whether the targets have been reached or not. Therefore, the development and application of criteria and indicators for measuring forest biodiversity is an important issue.

The scope of resource assessments follows the aims of international initiatives. Recently forest resources assessments have been extended towards the information on forest condition and "quality", rather than just on the resource itself. For instance, the TBFRA-2000 assessment was adapted to collect information on as many as possible of the Pan-European quantitative indicators on sustainable forest management (e.g. MCPFE 2000),

[1] Information sources and links to different international initiatives are provided in the references in Chapters 9 and 10. Regional and national initiatives are summarized for instance in "State of the European Forests and Forestry, 1999" (ECE/TIM/SP/16).

and there is an entire chapter dedicated to the Biological Diversity and Environmental Protection in the TBFRA-2000 Main Report (UN 2000).

In practice, international reporting has some limitations and poses some difficulties. To start, the data are aggregated at a relatively low level of resolution (commonly at national level) and summarise the various aspects of biodiversity into fairly few parameters. The selected indicators tend to represent a compromise from what is scientifically founded based on our current understanding, what kind of information is available and what can be agreed upon by the participating parties. Such a reporting is unlikely to serve as an early-warning system. For an evaluation of biodiversity the ecosystem composition, structure and processes need to be monitored at different scales ranging from continent and stand level to a scale of an organism, and from centuries and years down to months and days. The dimension and need for these different scales is acknowledged, while the current study concentrates on medium and large-scale elements relevant at regional or continental assessments.

Compilation of existing data for biodiversity assessments generally offers a large scope for bias and confusion (e.g. Vanclay 1998). The data have been collected in different countries according to different procedures, variables, definitions and standards during different time periods. Even the interpretation of traditional forestry variables needs special care in international efforts – and more so when assessing biodiversity-related variables.

The complete and long cycle of forest development, which incorporates various and varying stages each with different combinations of structural and compositional elements and each affected to a certain degree by anthropogenic and natural impacts, forms the rainbow of forest biodiversity, which the international assessments try to grasp. Therefore, only a few variables are unlikely to provide a comprehensive overview of the current situation, not even at European scale. Instead, an extended analysis by using credible and comparable data sources should improve our understanding and characterization of the variability of European forests.

Aims of the study

International reporting, large-scale assessments and long-term monitoring on forest biodiversity gain a lot of attention, but pose still difficulties to overcome. Some assessments concentrate on a few selected indicators, some on only a fraction of the geographical area and some use data, which are not directly comparable due to different definitions and standards used in national assessments. Joint analysis of the data of different initiatives has not always been possible, and political processes on the selection of criteria and indicators may have received attention instead of the plain assessment. Due to these reasons there is no comprehensive overview of the forest biodiversity and variety of the forests in Europe.

This paper aims to provide a comprehensive analysis of the forest biodiversity and variety of the forests in Europe. The analysis is carried out at continental and country level and geographically comprises the European countries covered by the TBFRA-2000. The study aims to establish a wide baseline assessment, which offers credible information on the current state also the possibility to detect and monitor changes in the future. This improved understanding of the variability of European forests should assist to select and refine indicators and monitoring systems. Subsequent to the assessment the paper aims to analyse the strengths, difficulties and deficiencies in current methods and data, and further to outline improvements and alternatives for the future biodiversity assessments and monitoring.

Both the Pan-European indicators on sustainable forest management (e.g. MCPFE 2000) and the framework of structural, compositional and functional key factors of European biodiversity as defined by Larsson et al. (2001)[2] are applied to provide a comprehensive and structured description of forest biodiversity. The paper does not repeat information, which has been directly provided either in the TBFRA-2000 Main Report (UN 2000) or in the follow-up report of the ministerial Conference the Pan-European (Third ministerial conference… 1998). Instead, it lists the availability and location of the indicators provided elsewhere and concentrates on analysing the different aspects of forest biodiversity.

[2] The key factors are one of the main results of the EC FAIR Project "Indicators for monitoring and evaluation of forest biodiversity in Europe (Bear)" CT97-3575.

The latest results and data from the Temperate and Boreal Forest Research Assessment (TBFRA-2000, UN 2000) have been available for the analysis, and additional material is further used to validate, complement and compare the results when appropriate. These data sources should allow the comparability between different European countries as far as possible due to the application of harmonised definitions and concepts. There will most likely never be "the perfect data" available for forest biodiversity assessment and monitoring in Europe, but the report tries to make the best out of the existing data, understanding their limitations.

Structure of the study

In the first part of the paper some aspects of the challenge of forest biodiversity monitoring are discussed (Chapter 2). Considering the wealth of the recent literature the discussion is merely a summary of the most essential premises considered in this work – and in no way tries to make up an all-inclusive review of the forest biodiversity monitoring. A preliminary list of key factors of European forest biodiversity as defined by Larsson et al. 2001 is also provided (Table 1).

Chapter 3 represents the frameworks of indicators or key factors applied in this study to describe, analyse and assess forest biodiversity. Firstly, the Pan-European indicators (MCPFE 2000) and subsequently the structural, compositional and functional key factors of European forest biodiversity (Larsson et al. 2001) are introduced. The availability and location of each indicator or key factor in the TBFRA-2000 Main Report is listed (Table 2 and 3). Furthermore, relationships between the Pan-European indicators on sustainable forest management and the key factors of European forest biodiversity are illustrated in Table 4.

The review of the structural, compositional and functional aspects of European forest biodiversity is the most extensive part of the paper (Chapters 4 - 6). Among the structural factors emphasis is given especially on the natural and protected forests and on structurally diverse forests (i.e. mixed species and uneven-aged forests) in Europe. Compositional key factors include a comparison of species numbers from different data sources, estimation of the importance of forests as a habitat for species and threatened species and a limited analysis of species richness per unit area.

In the discussion, conclusions are drawn of the value of the assessment. Furthermore, difficulties, improvements and options for the future assessment, monitoring and evaluation of forest biodiversity in Europe are discussed. The main attention is given to the variety of European forests, indicator framework, data concerns and scale issues. Finally, the management and monitoring for forest biodiversity protection and human-nature interactions are discussed.

2. The Challenge of Monitoring Forest Biodiversity

Biodiversity has been defined as "the totality of genes, species, and ecosystems in a region" (Rio 1992 Convention). It comprises the total variety of life on earth (Bibby et al. 1992) and includes not only the number of genes, species and ecosystems but also the diversity in their structure and function. Currently there is no unanimous definition for biodiversity – or for forest biodiversity. Instead, a recent study found 85 biodiversity definitions in use (De Long 1996).

Goals of the society

It is difficult to evaluate whether the state of forest biodiversity in Europe is good, bad, adequate or only 70% of what it should be. Similarly, it is difficult to answer the question often posed by forest industries - how much (forest protection, close-to-nature silviculture…) is enough. Ecological science is not in a position to decide why, and how much species diversity should be conserved. Such justifications should be rather found from the moral point of view (Thorne and Isermeyer 1997). Ultimately, the goals for forest protection or for the maintenance of forest biodiversity are set by the society, mostly as a complex compromise of different points of view.

Decisions for biodiversity protection are made in the environs of diverging aims, which are often not clearly defined, only partially quantifiable and due to the lack of common measurement scale difficult to compare. A similar scale for assessing the economic value of forestry activities in a rural community, the need for protected areas and the value of aesthetically beautiful forests is not trivial to discover – not to mention the competing other goals of the society. The following citation from the first Chapter "Facing up to our responsibilities" of the Biodiversity Action Plan from the European Commission (EC 2001) may well reflect this: "We have an ethical responsibility to preserve biodiversity for its intrinsic value. It also provides the food, fibre and drinks that our society needs. It is essential for maintaining the long-term viability of agriculture and fisheries, and is the basis of many industrial processes and the production of new medicines. It constitutes part of the world's natural capital on which many local communities and society at large relies. A loss of biodiversity is a loss of economic opportunity."

What research and development can do is to improve the understanding of ecological processes and of the linkages between ecològical, economical and social development. These assist in selecting appropriate criteria and indicators for describing biodiversity, in setting threshold values for the selected indicators, in developing monitoring methods and in improving the evaluation of different goals and decision making.

Relationship between biodiversity and "naturalness"

We seem to assume that biodiversity decreases with increasing intensity of ecosystem management, and that the increase in biodiversity automatically means an increase in ecological quality. In general, however, the human-induced landscape changes have had both negative and positive impacts on diversity. Some aspects of biodiversity may not be desirable, and diversity does not necessarily refer to concepts such as natural, stable or useful. When a natural landscape is fragmented, the overall community diversity may stay the same or even increase (Noss 1990). Instead of natural landscapes a patchwork of modified land types can be found (Niemelä 2000). Therefore, qualitative changes in community structure are often the best indicators of ecological disruption.

Species richness is a widely used and attempting variable to describe biological diversity in its simplicity – at first glace. The number of species, however, increases if we:

- Observe a large geographical area instead of a small one, e.g. if we look at a large country instead of a small one,
- Compare tropical rain forest with boreal forest ecosystem,
- Include species found in arboreta – but not in natural conditions – in the analysis or
- Increase the number of exotic species in the country.

For instance, in Germany the estimated number of exotic ferns and flowering plants amounts to 12,000, whereas the number of indigenous species is 3,000 (Anders and Hofmann 1997). As the traditional deciduous forest types have been replaced by pine forests in some areas of North-Eastern parts of Germany the species diversity has decreased, remained stable or increased, depending on the original situation (Anders and Hofmann 1997). Ammer (2000) found in studies in Bavaria, Germany that the species richness of fauna was largest in economically utilised mixed forests. In unutilised areas of natural beech or spruce forests the species richness was smaller. Furthermore, high species richness does not seem to coincide among different groups of organisms (Ferris and Humphrey 1999). The composition of different species may also indicate ecological disruption. For instance, generalist plants have extended their territory whereas especially indigenous specialist plants have been restrained in Eastern parts of Germany (Anders and Hofmann 1997).

The assessment of "naturalness" is often related to the concept of "biological integrity" (Angermeier and Karr 1994) or "ecosystem integrity" (De Leo and Levin 1997), which refers not only to biodiversity but also to the ability of an ecosystem to function and maintain itself. All in all, the definition of biological integrity is somewhat debated and vague, but essentially involves "...maintaining viable populations of native species, representation of ecosystem types across their natural range of variation, maintaining ecological processes, management over the long term, and accommodating human use within the above constraints" (Grumbine 1994). The concept includes the dynamic nature of the ecosystems, the maintenance of the structure and functioning typical to the natural habitats of a certain region and a sustainable relationship with the human component (De Leo and Levin 1997).

The international reporting is implicitly or explicitly directed at some aspects of biological integrity, even if it is not necessarily referred to in the documentation. Information such as forests and other wooded land by categories of naturalness, number of native species, percentage of endemic species, listings of the main invasive species or percentage of natural regeneration by native species collected in the TBFRA-2000 can be seen as targeting the concept of biological integrity and especially the native biodiversity. Among the Pan-European indicators such examples are changes in the area of natural and ancient semi-natural forest types, percentage of natural regeneration and changes in the proportions of stands managed for the conservation and utilisation of forest genetic resources with a differentiation between indigenous and introduced species.

Forest biodiversity as a result of quantity and quality

Forest biodiversity can be seen as a result of quantity and quality. The total forest area as such provides the potential for biological diversity, and it directly describes the area, which has remained free from agriculture and human settlements. It is also the most potential area, which can restore and enrich the forest biodiversity in the future. A large area can support more variation in adequate quantities than a small area. In contrast, however, a large area does not necessarily need to be biologically diverse and valuable - it can consist of exotic plantations, non-native species or be heavily polluted or fragmented into small separate patches. Thus, quantity needs to be complemented with quality.

Composition, structure and function have been distinguished and recognised as the primary elements of biological diversity and ecological integrity (Noss 1990). These components have also been applied by Larsson et al. (2001) in determining key factors and potential indicators for European forest biodiversity (Table 1) and could be interpreted as the determinants of the quality of the biodiversity. The components of structure, composition and function are highly interlinked with each other. For instance fire (functional factor) affects the species composition (compositional factor) and the age structure (structural factor), which again affect the susceptibility for certain disturbances (functional factor). As a further example, trees and stand structure affect birds (MacArthur and MacArthur 1961) or insects (Murdoch et al. 1972). Due to these inter-links and the fact that structures are often easier to quantify and monitor than the species or functions itself, structural diversity is a useful indicator not only on its own shake but also to provide indication of the other components.

Table 1. A preliminary list of key factors of European forest biodiversity (Larsson et al. 2001)

Scale	*Structural key factors*	*Compositional key factors*	*Functional key factors*
National or regional	Total area of forest with respect to – Forest types – Legal status/utilization or protection – Forest ownership – Tree species and age classes – Old growth or forest left for free development – Afforestation/deforestation	Native species Non-native or not "site-original tree species" Forest types	*For all scales:* Natural disturbance: fire, wind& snow, biological disturbance Human influence: forestry, agriculture and grazing, other land-use, pollution
Landscape	Number and type of habitats Continuity and connectivity of important habitats Fragmentation History of landscape use	Species with specific landscape scale requirements Non-native or not "site original" tree species	*For all scales:* Natural disturbance: fire, wind& snow, biological disturbance Human influence: forestry, agriculture and grazing, other land-use, pollution
Stand	Tree species ("site original", "not site original" and non-native) Stand size Edge characteristics (stand shape, ecotone, surrounding habitat) Forest history Habitat type(s) Tree stand structural complexity (horizontal and vertical) Dead wood (quality and amount) Litter (quality and amount)	Species with specific stand type and –scale requirements Biological soil conditions	*For all scales:* Natural disturbance: fire, wind& snow, biological disturbance Human influence: forestry, agriculture and grazing, other land-use, pollution

Description, analysis, assessment, evaluation and monitoring of forest biodiversity

There is sometimes confusion about the terms description, analysis, assessment and evaluation of the biodiversity. Description of the state of forest biodiversity in Europe, basically just provides an account of what the situation is like. Analysis is the process of considering the forest biodiversity carefully and in detail in order to understand it. Thus, an analysis should assist us to decide what are the targets we want to achieve, what are the means towards the desired direction and what are the indicators and methods, which at best assist us to describe and monitor the status of forest biodiversity. Assessment of a situation includes the consideration of all the facts, and an opinion of the position and of what is likely to happen. Evaluation is then a decision about the quality or value of forest biodiversity, based on the careful analysis of the different features. Evaluation is guided by the goals and values set by the society.

Biological diversity is rather a relative value than an absolute one and it can be evaluated by comparative studies. According to UNEP (1997) biodiversity assessment refers to an "analysis of the gap between the present state and the reference one". The reference state can be a certain point of time such as the year of the signature of the Convention on Biological Diversity, the situation before the industrial revolution or the time before any major human impact. The lack of reliable data is commonly an obstacle for the selection of early reference states. In some cases comparisons with similar ecosystems and habitats, which have been left out of the human activities may provide assistance in setting the reference state and targets and in understanding where we stand in terms of the loss of biodiversity. There are over 3,500 strictly protected forest reserves in European countries, and nearly 3 million ha of somewhat "natural" forests exist in protected areas in Europe (e.g. Parviainen et al. 2000). These form an important reference state for biodiversity studies. We may also compare different regions and countries to assess the areas, ecosystems or species, which are mostly endangered and to improve the understanding of the mechanisms, which lead to the depletion of biological diversity.

Even if there is a lack of a single definition of biodiversity, a lack of a unanimous list of indicators to describe forest biodiversity, a lack of comprehensive and "perfect" data to analyse the current situation of forest biodiversity in Europe and a lack of an agreed reference state or aims of management, we may meanwhile proceed by:

1. Agreeing on some criteria and indicators, which assist in describing the current state,
2. Recording and analysing the current status of these indicators,
3. Providing an indication of the desired direction, where the biodiversity situation should be developing in the (near) future,
4. Monitoring the changes in the indicator values regularly over time,
5. Continuously improving the understanding of underlying ecological processes, which support the natural biodiversity at different levels of the hierarchy, and the underlying socio-economic processes which lead to depletion of natural biodiversity,
6. Assisting policy making in the follow-up of the developments and in the improvement of selected criteria and indicators, monitoring, and in the setting up of new targets as new knowledge becomes available.

3. Indicators and Key Factors for the Description of Forest Biodiversity

Concrete parameters or indicators are needed to describe, assess and evaluate the current state of forest biodiversity, and to monitor changes in the future. It would be surprising to find a common agreement on only a few indicators after creating such a large number of different definitions. Thus, also the number of suggested indicators is large. This is partly justified, because the characterisation of the biodiversity in a forest ecosystem can only partly be achieved by the simple compilation of lists of species or habitats (Jeffers 1996). "If diversity could be encapsulated into one single figure, it would be closer to simplicity and uniqueness of behaviour, rather than to complexity and diversity of behaviour, as it is often implicitly assumed (Franc 1998)." Besides the need for a comprehensive description of a complex issues, the large number of proposed indicators also originates from the development of several potential indicators within different disciplines and initiatives. It is likely that the improved understanding of the biodiversity issues refines the selection of the most appropriate indicators.

Two main sets of biodiversity measures developed for European forests are applied in this study, the Pan-European indicators (e.g. MCPFE 2000) and the key factors for European forest biodiversity (Larsson et al. 2001). The emphasis is given to the latter. These sets and their relationships are reviewed shortly in the following two chapters.

3.1 Pan-European Indicators for Sustainable Forest Management

The Pan-European process, i.e. Ministerial Conference on the Protection of Forests in Europe, started in Strasbourg in 1990 and has been followed by the conferences in Helsinki (1993) and Lisbon (1998). The resolutions of the process have been signed by 37 European countries. Resolution H2 "General guidelines for the conservation of biodiversity of European forest" was adopted at the Helsinki Conference. The Resolution L2 adopted at the Lisbon Conference further defines "Pan-European criteria, indicators and operational level guidelines for sustainable forest management". These comprise a set of six criteria and 27 indicators, 20 of which are quantitative, for the sustainable forest management in Europe. The six criteria deal with the maintenance and appropriate enhancement of:

- Forest resources (Criterion 1),
- Forest ecosystem health and vitality (Criterion 2),
- Productive functions (Criterion 3),
- Biological diversity (Criterion 4),
- Protective functions (Criterion 5) and
- Other socio-economic functions and conditions (Criterion 6).

The TBFRA-2000 was adapted to collect information on as many as possible of the Pan-European quantitative indicators. It provides information directly on eight of the twenty quantitative Pan-European indicators on sustainable forest management, five further indicators are covered partly or through data modifications and there is no information related to seven of the quantitative indicators (Table 2). In the follow-up report of the ministerial Conference the Pan-European indicators have been reported in tables and graphs together with further national documentation (Third ministerial conference… 1998). The reported parameters correspond to the state of the TBFRA analysis and reporting by March 1998, prior to the final validation and analysis. The information in the final form can be found in the TBFRA-2000 Main Report.

Table 2 lists the Pan-European indicators and their availability as well as their location in the TBFRA-2000 Main Report. They are shortly summarised in the following.

Indicators reported directly in the TBFRA-2000

The indicators, which are directly available from the TBFRA-2000 Main Report are the ones related to

- forest area,
- carbon storage,
- changes in serious defoliation of forests,
- balance between growth and removals,
- proportion of forest area managed according to the management plan,
- proportion of natural regeneration and
- forest area managed primarily for soil protection and provision of recreation.

Within this information there are some minor differences in definitions. Firstly, according to the Pan-European indicators the balance between growth and removals refers to the past ten years. The time period in the TBFRA-2000 has not necessarily been 10 years and it has not been the same for all the countries. Secondly, the term managed may be understood in several ways and is not necessarily easy to define. The definition used in the TBFRA-2000 assessment is: "Forest and other wooded land which is managed in accordance with a formal or an informal plan applied regularly over a sufficiently long period (5 years or more). The management operations include the tasks to be accomplished in individual forest stands (e.g. compartments) during the given period." Thirdly, the methods and statistical coefficients used to transform the stem wood volume into total biomass of all tree parts and, finally, into carbon have been described in the Chapter III of the TBFRA-2000 Main Report.

Indicators available to a certain degree in the TBFRA-2000

The Pan-European indicators almost systematically involve data on the change of the indicator value during a certain time period besides the current state. In contrast, estimates of change have been reported only for a small number of key parameters in the TBFRA-2000 – due to the severe methodological difficulties related to monitoring. In most cases the definitions of the variables have been changed or the information has been collected for the first time so that change detection is not viable. Pan-European indicators partly covered by the TBFRA-2000 information content are as follows:

- Changes in the timber resources. The changes in the total growing stock have been reported in the TBFRA-2000 process (1.2.a) and the changes in the mean volume of the growing stock on forest land (1.2.b) are somewhat available through the change of the forest area and the change of the total growing stock. There is, however, no information concerning the change of the age structure or diameter distribution of the forests (1.2.c). Instead, the current state is reported.
- Serious damage caused by biotic or abiotic agents. The indicator comprises the extent of damaged caused by four different agents. Three of these are directly available from the TBFRA-2000 main report and have been supplied by the ICP-forests (EC and UNECE 1999a,b). The fourth one is the percentage of regeneration area seriously damaged by game or other animals or by grazing. The information collected in the TBFRA-2000 comprises the whole forest and wooded land area damaged by wildlife and grazing – and not only the regeneration area.
- Changes in the natural and protected forests. Current state is available.
- Changes in the number and proportion of threatened species. Current state is available.
- Changes in the proportions of mixed stands. Current state is available.

Indicators for which additional data sources are needed

Seven of the Pan-European indicators are not available in the report, namely deposits of air pollutants, changes in the nutrient balance and acidity, information on non-wood forest products, conservation and utilisation of forest genetic resources, forests managed for water protection, size of the forest sector and forestry employment. Information on these factors can be found from the national reports of the follow-up documentation of the ministerial conferences (Third ministerial conference... 1998). In the following, the data availability and sources in Europe are shortly reviewed:

- Deposits of air pollutants and changes in the nutrient balance and acidity are monitored in Europe through the International Co-operative Programme on Assessment and Monitoring of Air Pollution Effects on Forests of the UNECE (ICP Forests) (EC and UNECE 1999a,b). The intensive monitoring programme (Level II) based currently on about 860 plots around Europe studies especially these three factors. To date, only a limited number of plots have been analysed and there is limited information on changes.
- A number of countries have listed information concerning non-wood products such as mushrooms and berries, medicinal and herbal plants, decorative foliage, fodder and forage, hunting and game products and other non-wood products (Chapter VI). Some quantities have been reported in the follow-up report of the ministerial Conference (Third ministerial conference... 1998). The data on the quantity of the non-wood products are, however, limited, often partial, collected by using different nomenclatures and seldom collected based on statistically designed inventories. Reliable comparisons between countries are close to impossible. Clearer definitions of the products and further information on the source of the estimates are needed in the future. There is an on-going EU FAIR-project Scale-dependent monitoring of non-timber forest resources based on indicators assessed in various data sources (MNTFR), which will provide assistance for future assessments.
- The information related to the size of the forestry sector and forestry employment is not difficult to obtain from general statistical sources (Eurostat and national statistical agencies). Attention should be paid to the compatibility of the definitions, especially between the European Union and other European countries.
- Information on the conservation and utilisation of forest genetic resources as well as on the forests managed for water protection is available to some degree from the national reports in the follow-up report of the ministerial Conference (Third ministerial conference... 1998).

Table 2. Quantitative Pan-European indicators for sustainable forest management (MCPFE 1998) and the corresponding data in the TBFRA-2000

PAN-EUROPEAN CRITERIA AND INDICATORS FOR SUSTAINABLE FOREST MANAGEMENT	*INFORMATION PROVIDED BY TBFRA-2000*			
	Indicator availability[3]			
Description of quantitative indicators	Directly	Through Modifications/ no estimate on change	Not available	*Location of the corresponding data (TBFRA-2000 Main Report)*
CRITERION 1: FOREST RESOURCES				
1.1. Area of forest and other wooded land and changes in area (classified, if appropriate, according to forest and vegetation type, ownership structure, age structure, origin of forest)	x			Chapter 1: Area of forest and other wooded land: status and changes Chapter 2: Ownership structure Chapter 3: Amount of uneven-aged and even-aged forests (Tables 25-28) as well as aged-class distribution (Tables 29-32)
1.2. Changes in				
a) total volume of growing stock	x			a) Table 37
b) mean volume of the growing stock on forest land (Classified, if appropriate, according to vegetation zones/site classes)		x		b) Change in the forest area (Table 7) and change in growing stock (Table 37)
c) age structure or appropriate diameter distribution classes		x		c) Current state in Tables 29-32
1.3. Total carbon storage and, a change in the storage in forest stands	x			Annex 3B.2 The carbon store of woody biomass Annex 3B.3 The carbon balance of woody biomass on forest and other wooded land
CRITERION 2: HEALTH AND VITALITY				
2.1. Total amount of and, changes over the past 5 years in depositions of air pollutants (assessed in permanent plots)			x	No information. Estimation of the area damaged by known local pollution sources given in Table 70.

[3] 'Available directly' means that the given indicator is available directly in the defined form of the TBFRA-2000 Main Report. 'Through modifications' means that the information can be obtained by modifying the data collected in the TBFRA-2000 process. If the indicator is 'not available', the information closest to the content of the indicator has been listed.

2.2. Changes in serious defoliation of forests using the UNECE and EU defoliation classification (Classes 2, 3 and 4) over the past 5 years)	x			Tables 76-78: All species and conifers and broadleaves separately
2.3. Serious damage caused by biotic or abiotic agents:				
a) severe damage caused by insects and diseases with a measurement of seriousness of the damage as a function of (mortality or) loss of growth	x			a) Table 70
b) annual area of burnt forest and other wooded land	x			b) Table 70, Tables 72-75 offer further detailed information on forest fires
c) annual area affected by storm damage and volume harvested from these areas	x			c) Table 70
d) proportion of regeneration area seriously damaged by game and other animals or by grazing			x	d) Area of forest and other wooded land damaged by wildlife and grazing in Table 70
2.4. Changes in nutrient balance and acidity over the past 10 years (pH and CE); level of saturation of CEC on the plots of the European network or of an equivalent national network			X	No information
CRITERION 3: PRODUCTIVE FUNCTIONS OF FORESTS				
3.1. Balance between growth and removals over past 10 years	X			Table 52: Fellings as percent of net annual increment and removals as percent of fellings on forest available for wood supply Table 49: Annual removals overbark on forest and Table 42: Gross and net annual increment The time period is not necessarily the past 10 years–and not the same for all countries
3.2. Percentage of forest area managed according to management plan or management guidelines	X			Table 10: Percentage of managed forest and other woodland of the total. Management "...in accordance with a formal or informal plan applied regularly..."
3.3. Total amount of and changes in the value and/or quantity of non-wood forest products (e.g., hunting and game, cork, berries, mushrooms, etc.)			X	No information on the amounts in the TBFRA-2000 Main Report. Some partial data are available in the follow-up report of the Helsinki-process (Third ministerial conference... 1998)

CRITERION 4: BIOLOGICAL DIVERSITY				
4.1. Changes in the area of				No estimate on the change, current state of:
a) natural and ancient seminatural forest types		x		a) The amount of forest and OWL by categories of "naturalness" (Tables 53-54)
b) strictly protected forest reserves		x		b) Forest and OWL in IUCN protection categories I to VI (Tables 55)
c) forests protected by special management regime			x	c) No information
4.2. Changes in the number and percentage of threatened species in relation to total number of forest species (using reference lists e.g., IUCN, Council of Europe or the EU Habitat Directive)		x		No estimate on the change, current state of the total and endangered numbers of forest occurring species, separately assessed for trees, vascular plants other than trees, ferns, mosses, lichens, mammals, birds, other vertebrates and butterflies & moths (Tables 56-64).
4.3. Changes in the proportions of stands managed for the conservation and utilisation of forest genetic resources (gene reserve forests, seed collection stands, etc.); differentiation between indigenous and introduced species			X	No direct information. Instead, information of usage of original/introduced species in forest regeneration, extension and natural colonization are available (Tables 65-68). The origin of the planting material, indigenous or introduced, from known local or non-local or unknown local provenances (Table 69).
4.4. Changes in the proportions of mixed stands of 2-3 species		x		No estimate on change, current situation in Table 4
4.5. In relation to total area regenerated, proportions of annual area of natural regeneration	X			Table 68
CRITERION 5: PROTECTIVE FUNCTIONS				
5.1. Proportion of forest area managed primarily for soil protection	x			Table 79
5.2. Proportion of forest area managed primarily for water protection			x	No information
CRITERION 6: OTHER SOCIO-ECONOMIC FUNCTIONS AND CONDITIONS				
6.1. Share of the forest sector from the gross national product			x	No information
6.2. Provision of recreation: area of forest with access per inhabitant, % of total forest area	x			Table 81
6.3. Changes in the rate of employment in forestry			x	No information

3.2 Structural, Compositional and Functional Key Factors of Forest Biodiversity

The political Pan-European process has been supported by considerable research effort targeting the implementation of the political process in forest resources management. Major work for defining forest biodiversity indicators for country, landscape and stand level has been carried out in the EU Commission FAIR Project "Indicators for monitoring and evaluation of forest biodiversity in Europe (Bear)". The draft report of the project has been available for this study. Therefore, some minor changes and improvements are likely to be found in the final printed report (Larsson et al. 2001).

The Bear project agreed on a single common scheme of key factors relevant to the biodiversity of all European forests (see Table 1). These factors consist of structural, compositional and functional elements of European forest biodiversity[4]. They are adapted and applied in this report to describe the current status of forest biodiversity. Only the national level key factors (and the corresponding gross list of potential indicators) are relevant within this context. The definitions and nomenclature of the potential indicators are understood in a flexible manner, because so far, no clear recommendations or priority lists have been proposed and also because no final definitions have been agreed upon. In that way the interpretation of the information content is not as rigorous as in the case of the Pan-European indicators, where a clear set of indicators exists.

The relative importance of different biodiversity elements varies considerably in different European forests (Larsson et al. 2001). Therefore, the Bear-project identified about 30 different forest types to assist the adaptation of the key factors to operational stand and landscape level management and monitoring. These landscape and stand level approaches, however, have not been tested or applied in this study.

The availability of the TBFRA-2000 data concerning a particular key factor or potential indicator is listed in Table 3. At the national level the TBFRA-2000 provides information covering almost all of the structural, compositional and functional categories – at least to some extent. Only some deficits can be identified. The lack of information available on change was already stated in connection with the Pan-European indicators. Further, information on the extent of "real" forest types such as those identified by Larsson et al. (2001) is missing from the TBFRA-20000 results so that only the extent of natural and semi-natural forests as well as plantations can be separated. The data on the extent of non-native or not "site-original" species are available only from the regeneration areas. In the TBFRA-2000 Main Report there is also fairly little information available on other land use, agriculture and grazing and on the "specific" forms of silviculture. Despite these weaknesses the TBFRA-2000 provides a considerable amount of data on the key factors of the European forest biodiversity.

[4] For instance Jones and Riddle (1996) divide biodiversity indicators into species metrics and integrated metrics. Species metrics include for instance species diversity and richness, and integrated metrics evaluate the composition, structure and function of biodiversity. As understood by Larsson et al. (2001) the compositional key factor encompasses also the species metrics and data.

Table 3. National level indicators for describing European forest biodiversity (adapted from Larsson et al. 2001) and the corresponding data in the TBFRA-2000

BEAR-PROJECT RECOMMENDATIONS (Larsson et al. 2001)	LOCATION OF THE CORRESPONDING DATA (TBFRA-2000 Main Report)
Structural key factors – country level	
Total area of forest with respect to	Table 1: Total area of forest and other wooded land (OWL) Table 2: % of forest and/or OWL of land area
• Forest types	No direct information of forest types as understood by the forest management. Tables 53&54: Forest and other wooded land by categories of "naturalness": undisturbed by man, semi-natural, plantations
• Legal status/utilization or protection	Table 55: Forest and OWL in IUCN protection categories I-VI
• Forest ownership	Tables 9-24
• Productive forest	Table 5: Forest available for wood supply. The terms forest available and not available for wood supply have replaced productive and unproductive forest used in previous assessment in order to improve and clarify the definitions.
• Tree species	Table 3& 4: Forest and OWL by species groups • Predominantly coniferous • Predominantly broadleaved • Predominantly bamboos, palms, etc. • Mixed
• Age classes	Tables 25-28: Uneven-aged and even-aged high forest available for wood supply: total and in species groups Tables 29-31: Age-class distribution of even-ages high forest available for wood supply: total and in species groups. Three age classes have been distinguished, namely forests <40 years, 40 - 80 years and forests >80 years of age.
• Old growth/forest left for free development	Table 6: Forest not available for wood supply Table 55: Protection status Tables 53&54: Forest and other wooded land by categories of "naturalness": undisturbed by man, semi-natural, plantations
• Afforestation/deforestation	Table 65: Annual averages of area of extension of forest and natural colonization of other wooded land over ten-year period
Compositional key factors – country level	
Native species	
• Red-listed species	Tables 55-64: Number of endangered species of Trees, Vascular plants other than trees, Ferns, Mosses, Lichens, Mammals, birds, other vertebrates and butterflies&moths.
• Other forest species	Tables 55-64: Reported number of species: Trees, Vascular plants other than trees, Ferns, Mosses, Lichens, Mammals, birds, other vertebrates and butterflies&moths.
Non-native or not "site-original" tree and other species	Table 65: Area (percentage) of introduced species used in forest regeneration, extension and natural colonization annual averages of area Table 66: Ares (percentage) of introduced species used in different types of forest regeneration, annual averages Table 69: Origin of planting material used in forest

Functional key factors – country level	
Natural Influence	
• Fire	Table 71: Area of damage to forest and other wooded land separated by known causes Tables 72-75: Forest fires: number, total area burned, area of forest/OWL burned
• Wind and snow	Table 71: Area of damage to forest and other wooded land/storm, wind, snow or other identifiable abiotic factors
• Biological disturbance (incl. Pests)	Table 71: Area of damage to forest and other wooded land • insects and diseases • wildlife and grazing
Anthropogenic influence	
• Forestry, "normal" silviculture	Table 16: Proportion of managed forests from the total Table 5: Forest available for wood supply (ha) by silvicultural categories high forest and coppice Table 48: Annual fellings overbark on forest available for wood supply by species groups Table 52: Fellings per hectare and fellings as percent of net annual increment Table 65: Annual averages of area of regeneration, extension and natural colonization Table 68: Types of regeneration and extension of forest
• Forestry, "specific" silviculture	Table 79: area where forest and other wooded land is managed primarily for soil protection
• Agriculture and grazing	
• Other land use	Table 81: Area of forest and OWL where access to public is legally not allowed
• Pollution	Table 70: Area of damage to forest and other wooded land/known local pollution sources Tables 76-78: Amount and changes in serious defoliation of forests

3.3 Relation between the Key Factors and Pan-European Indicators

Only one of the six criteria for Pan-European sustainable forest management explicitly deals with biodiversity, namely Criterion 4: "Maintenance, conservation and appropriate enhancement of biological diversity in forest ecosystems". It includes five quantitative indicators (see Table 2). The extent of the key factors of European biodiversity as understood by Larsson et al. (2001) is broader. Table 4 illustrates the relationships between the structural, compositional and functional key factors of European forest biodiversity and the Pan-European indicators for sustainable forest management. The key factors have been interpreted relatively broadly, and as a result not only the Pan-European Criterion 4 deals with biodiversity – but also some further aspects and indicators of sustainable forest management. These key factors are seen in this paper as complementary to the Pan-European indicators in that they offer a wider and structured scope for describing the variety of European forests.

Table 4. Relationships between the quantitative Pan-European indicators for sustainable forest management and the key factors of European forest biodiversity

Quantitative Pan-European indicators for sustainable forest management (MCPFE 1998)	*Key factors of European forest biodiversity (Broad interpretation of Larsson et al. 2001)*
CRITERION 1: FOREST RESOURCES	
1.1. Area of forest and other wooded land and changes in area	Structural factor
1.2. Changes in timber resources	-
1.3. Total carbon storage and balance	-
CRITERION 2: HEALTH AND VITALITY	
2.1. Amount and in depositions of air pollutants	Pollution as such considered as a functional factor (anthropogenic influence)
2.2. Changes in serious defoliation of forests	Pollution as such considered as a functional factor (anthropogenic influence)
2.3. Serious damage caused by biotic or abiotic agents	Functional factors (natural influence)
2.4. Changes in nutrient balance and acidity	Pollution considered as a functional factor (anthropogenic influence)
CRITERION 3: PRODUCTIVE FUNCTIONS OF FORESTS	
3.1. Balance between growth and removals over past 10 years	Functional factor (anthropogenic influence, forestry)
3.2. Percentage of forest area managed according to management plan or management guidelines	Functional factor (anthropogenic influence, forestry)
3.3. Value and/or quantity of non-wood forest products	-
CRITERION 4: BIOLOGICAL DIVERSITY	
4.1. Changes in the areas of a) natural and ancient seminatural forest types b) strictly protected forest reserves c) forests protected by special management regime	 a) structural factor b) structural factor c) Combination of structural and functional factor?
4.2. Changes in the number and percentage of threatened species	Compositional factor
4.3. Changes in the proportions of stands managed for the conservation and utilisation of forest genetic resources; differentiation between indigenous and introduced species	Structural factor
4.4. Changes in the proportions of mixed stands of 2-3 species	Structural factor
4.5. Proportions of annual area of natural regeneration	Structural factor
CRITERION 5: PROTECTIVE FUNCTIONS	
5.1. Proportion of forest area managed primarily for soil protection	Functional factor - anthropogenic influence (other land use)
5.2. Proportion of forest area managed primarily for water protection	Functional factor - anthropogenic influence (other land use)
CRITERION 6: OTHER SOCIO-ECONOMIC FUNCTIONS AND CONDITIONS	
6.1. Share of the forest sector from the gross national product	-
6.2. Provision of recreation	Functional factor– anthropogenic influence (other land use)
6.3. Changes in the rate of employment in forestry, notably in rural areas	-

4. Structural Aspects of Forest Biodiversity in Europe

At national or regional levels the structural component of forest biodiversity at best describes the overall framework and pattern of the forests. Estimation of the proportions of exploited forests or forests available for wood supply, managed forests, afforestation, deforestation, age structure and plantations describe the structure, which has been modified by human impact. Estimates of protected forests and forests left for natural development illustrate the potential still available for the maintenance of original structures, composition and functioning of the ecosystem. Information on mixed species or un-even aged forests describes structurally diverse forests.

4.1 Area of Forests and Other Wooded Land

The area of forests and other wooded land provides the potential for forest biodiversity. Within Europe there is a contrast between the relatively high cover of forest and other wooded land in the Nordic countries and the Iberian peninsular (59% and 50% respectively) compared with the low cover in North-Western Europe (11%) (TBFRA-2000 Main Report). Other wooded land increases the forest area considerably especially in the countries around the Mediterranean (Figure 1). On average, the proportion of forest area as a percentage of forest plus other wooded land is 87 %. During the last decade there has been an average annual increase in forest area by about 500,000 ha, and at the same time, a decrease in the area of other wooded land by about 200,000 ha in Europe.

Even if shrinking of the extent of natural and semi-natural habitats is seen as one of the greatest losses of biodiversity (e.g. UNEP 1999), forest area as a total is not decreasing in Europe. Instead, human impact results in fragmentation and homogenisation in the landscape (Niemelä 2000). Therefore, the qualitative aspects of forests should be understood and evaluated in order to assess the loss of biodiversity.

Figure 1. Relationship between the area of forests and other wooded land in European countries[5]. Each dot represents one country. Data: TBFRA-2000.

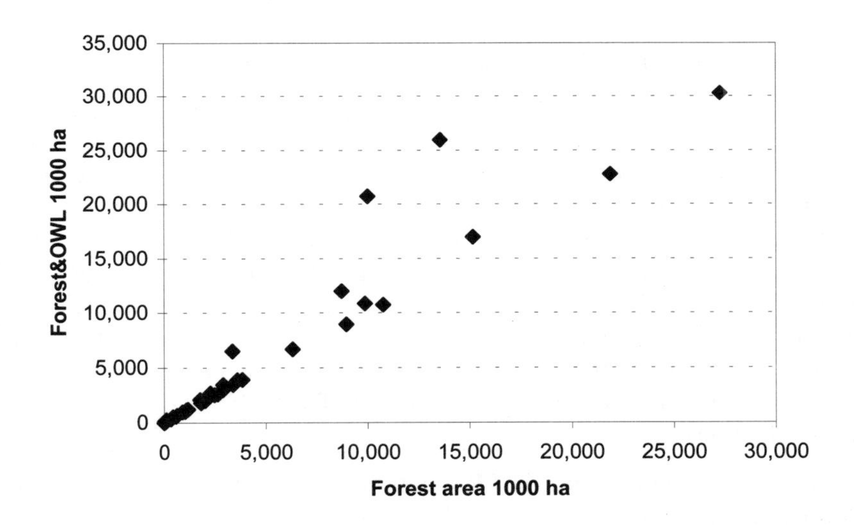

[5] Positive deviations above the drawn line indicate that the total wooded area includes not only dense forests but also shrub or bush cover or sparse stocking levels.

4.2 Natural and Protected Forests

"Naturalness" in European forests

Forests are moderately disturbed compared with agricultural fields, urban parks or gardens - despite of the long human influence in Europe. The signs of human interaction with nature in the landscape are varied, contrasting and localised. The human-induced landscape changes have had both negative and positive impacts on forest diversity. Over centuries humans have created new habitats by opening and diversifying the landscape. For instance in boreal landscapes, the majority of current deciduous forests has been found on former agricultural lands (Mikusiński and Angelstam 1998). The pattern was explained as a result of unintended succession of abandoned fields, less intensive forest management in privately owned forests and of the aesthetic value of deciduous trees close to human settlements. Some of these landscapes, created by man, belong nowadays to protected areas.

The existence of natural and protected forests is crucial for the protection of natural processes and species related to the natural process, for understanding the ecological principals, and for reference when setting up management priorities and plans (Parviainen et al. 2000). Natural forests also serve as model for nature-oriented silviculture.

The determination of the concept and degree of "naturalness" is not simple (e.g. TBFRA-2000 Main Report). Generally, naturalness[6] means forests, which have not been affected by the man. In Europe the definition is problematic, since due to the dense population forests have been subjected to influence of humans widely and for a long time. In Southern, Atlantic and Central parts of Europe the forest structure was changed due to human impact by the Middle Ages and in Northern Europe during the last 300 - 400 years (Bengtsson et al. 2000).

Austria was the first European country to carry out a special inventory to assess the degree of naturalness of its forests (Grabherr 1997). Eleven criteria of naturalness were applied, among them the naturalness of tree species composition and that of ground flora, intensity of utilisation and the amount and quality of dead wood. In general, however, it is difficult to summarize the area of natural forests in Europe due to different policies and degrees of forest protection in different countries (see Parviainen et al. 1999).

Extent of natural, semi-natural and plantation forests in Europe

There are hardly any truly natural forests remaining in Europe, excluding some remote almost inaccessible areas or areas where extreme climatic or topographic conditions prevail (e.g. Kuusela 1994). Only about 4 % of the forest area has been classified as undisturbed by man in the TBFRA-2000 Main Report (Figure 2). Bücking et al. (2000) report similar results for virgin forests in Europe. They understand virgin forests as areas, which have been continually forested since the conditions have been suitable for tree growth, i.e. since the end of the Ice Age (Central and Northern Europe) or since the Tertiary period (the Mediterranean). The amount of virgin forests is very small in Europe, and they certainly exist in the Scandinavian countries, in Karelia, Archangelsk and the Komi Republic of the Russian Federation, and in the Alps and the Balkan region. Some further estimates state that only 0.2% of the Central European deciduous forests have remained in a somewhat natural state (Hannah et al. 1995) and that there are nearly 3 million ha of somewhat "natural" forests in protected areas in Europe (Parviainen et al. 2000).

Only 12 countries out of 38 report in the TBFRA-2000 that the percentage of forests undisturbed by man is larger than 1%: Among these are the EU-countries Sweden (16.1 %), Finland (5.8 %) and Portugal (1.6 %). Generally the countries are located in the northern (Sweden, Norway and Finland) or eastern (Albania, Bulgaria, Poland, Slovenia, Slovakia, Romania and Turkey) Europe.

[6] Examples of other concepts close to naturalness of forests are native, ancient woodland, virgin forest, old-growth forest, primary forest and old forest (Peterken 1997).

The largest areas of forest plantations can be found in Spain, Turkey, the United Kingdom, Bulgaria, France, Portugal and Ireland (Figure 3). No plantations or less than 1000 ha were reported in Austria, Czech Republic, Finland, Germany, Slovenia and Switzerland. The plantations have been defined as follows (TBFRA-2000):

> *"Forest stands established by planting or/and seeding in the process of afforestation or reforestation. They are either of introduced species (all planted stands), or intensively managed stands of indigenous species which meet all the following criteria: one or two species at plantation, even age class, regular spacing.*
>
> *Excludes: Stands which were established as plantations but which have been without intensive management for a significant period of time. These should be considered semi-natural."*

Especially the exclusion of stands, which were established as plantations but which have been without intensive management for a significant period of time, allows different interpretations by countries. For instance, Germany has not reported any plantations even though the majority of the large afforestations after the First and Second World War was carried out by planting or seeding. Especially the genetic uniformity caused by the planting material still affects today.

Figure 2. Percentage of plantations, semi-natural forests and forests undisturbed by man in Europe. Data: TBFRA-2000.

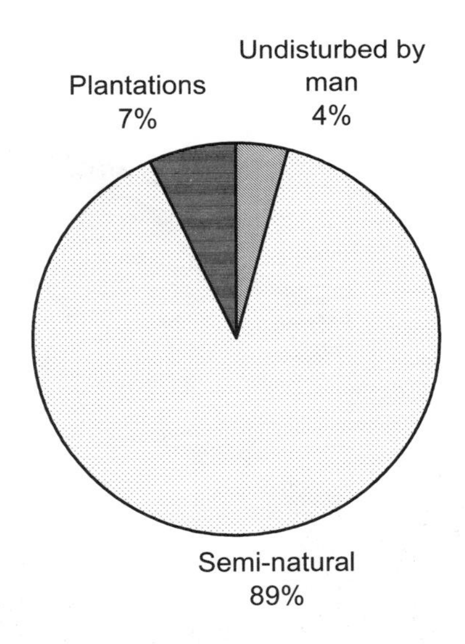

Figure 3. Plantation area in different European countries. Data: TBFRA-2000.

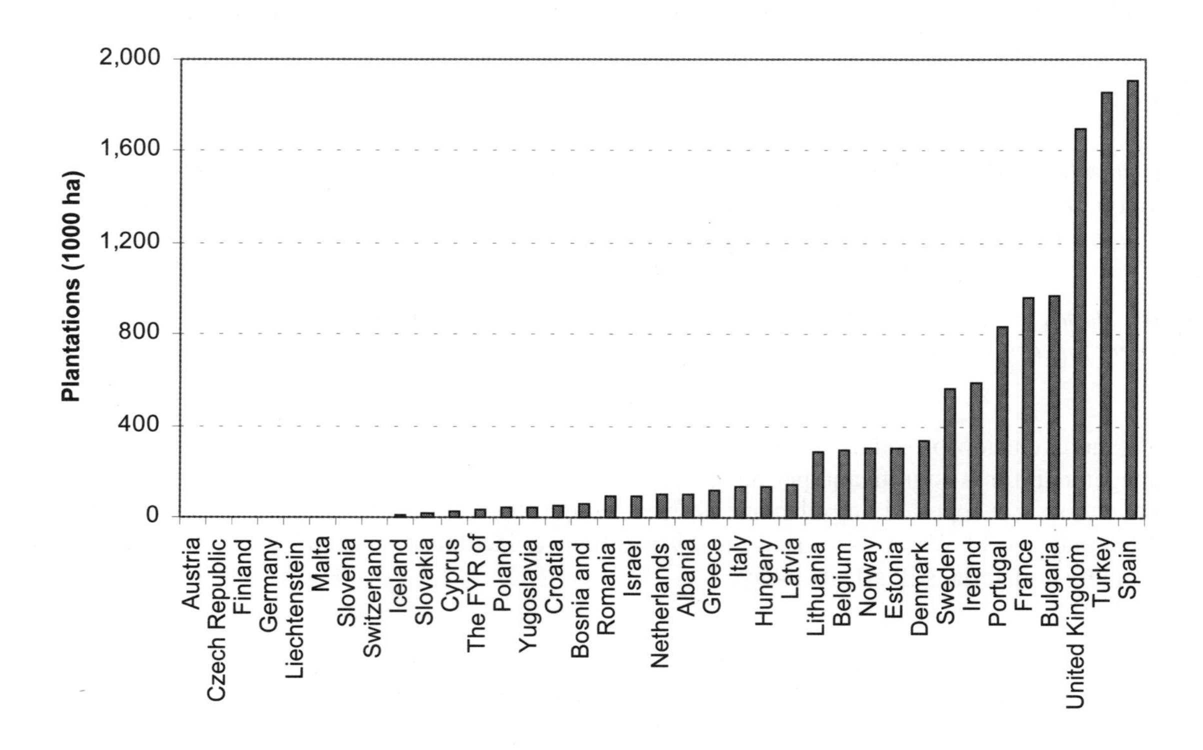

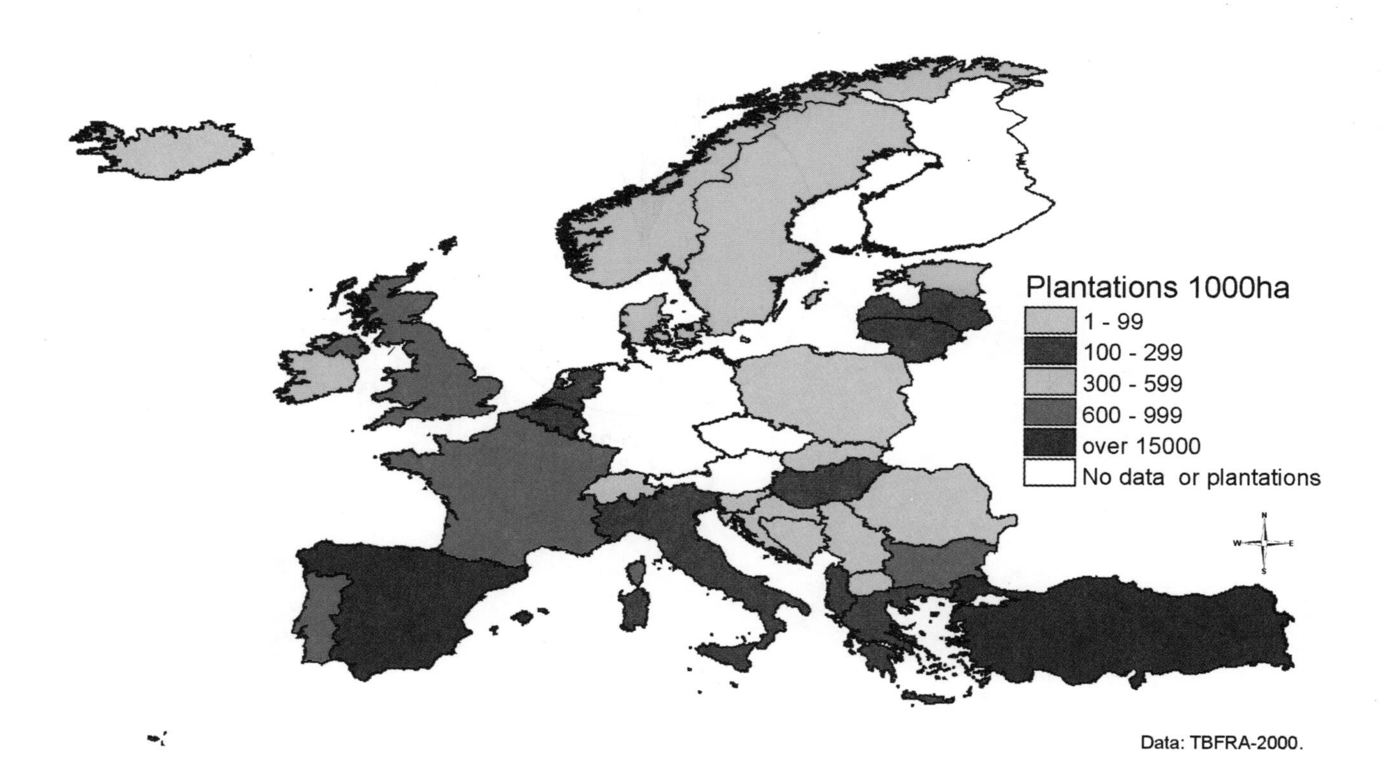

Variety in forest protection in Europe

The protection status of a forest is often related to forest condition so that protected areas tend to be in a better state than exploited ones (e.g. Kapos and Iremonger 1998). Protected forests are also likely to be closer to the "natural state" than the exploited ones. The forest protection concept in Europe is more versatile than for instance that of America. Strict reserves, which are left out of human impact, are strongly complemented with more open protection categories and close-to-nature approaches of silviculture, especially in densely populated Central European countries. Two further lines in the objectives of forest protection can also be distinguished (Parviainen et al. 2000). Firstly, forests can be protected mainly for the conservation of biodiversity and natural ecosystem processes. In this case the main aim of the protection is the forest itself. Secondly, forests can be protected in order to protect soils and ground water, or to protect areas against avalanches and other catastrophes. In these cases forests serve as a mean to achieve the ultimate goal of protection.

At the Fourth World Congress on National Parks and Protected Areas, in Caracas in 1992, IUCN/World Conservation Union suggested that around 10% of the Earth's land surface should be in protected areas to conserve a full complement of biodiversity. This target was repeated in 1995 in a joint IUCN/WWF forest strategy. The situation of forest protection in Europe, however, is in no way simple and clear.

There are nearly 90 different categories of protected forests ranging from national parks to aesthetic forests, and the forest protection policy and implementation differ widely between European countries (Parviainen et al. 2000). Strict forest reserves, for instance, may be protected via Acts or laws related to forestry, nature conservation or to both, or via administrative regulations, ministerial edicts or private contracts, depending on the country (Bücking et al. 2000). There are also privately protected unmanaged areas, the extent of which is difficult to assess. The qualitative order of the protective areas is, furthermore, not unambiguous. In some countries national parks are considered to be of higher "quality" than strict forest reserves, whereas in the IUCN-classification national parks belong to category II instead of the strict reserves category I. In many countries national parks also include smaller, strictly protected core areas, which are left for free development (Bücking et al. 2000). Bücking et al. (2000) further state that often the Central European national parks are effectively speaking strict forest reserves surrounded by managed protected areas. This applies to Austria, France, Finland, Germany, Greece, Italy, Norway, Portugal, Slovenia, Spain and Sweden.

Protection categories applied by IUCN and COST Action E4: "Forest Reserves Research Network"

The definition used in the TBFRA-2000 for protected areas is based on the IUCN definition – i.e. "*an area of land and/or sea especially dedicated to the protection and maintenance of biological diversity, and of natural and associated cultural resources, and managed through legal or other effective means*". Information related to protected forests has been collected on:

- Forests not available for wood supply for conservation reasons,
- Forests and other wooded land under strict protection (as defined by IUCN categories I and II),
- Forests and other wooded land under more flexible forms of protection (i.e. forest and woodland in IUCN categories III-IV).

There has been considerable confusion and disagreement on the interpretation of the protection categories. Therefore, additional information has been sought after to demonstrate the current state of protected forests.

The COST Action E4: 'Forest Reserves Research Network' was the first systematic attempt in Europe to create a network on forest reserves and to collect information on strictly protected forests (Parviainen 2000). The action concentrated especially on forests protected for the conservation of biodiversity and natural processes within forest ecosystems. Twenty-seven countries participated in 1995 - 1999. One of the main findings of the action was that harmonising and analysing the protected forest categories was much more difficult than originally planned, due to the variable and ambiguous definitions used in different countries[7]. The definitions applied to different categories of forest protection in the TBFRA and COST Action E4 are summarized in Table 5.

---

[7] The participants also called for co-operation between the COST Action E4 experts, TBFRA correspondents and the IUCN national representatives to better integrate protected forest data with other forest resources information (Parviainen et al. 2000).

Table 5. Some definitions related to international forest protection categories in Europe

Term	*Process*	*Definition*
Forest not available for wood supply due to protection reasons	TBFRA-2000	Forest where legal or specific environmental restrictions prevent any significant supply of wood. Includes forests with legal restrictions or restrictions resulting from other political decisions, which totally exclude or severely limit wood supply, inter alia for reasons of environmental or biodiversity conservation, e.g. protection forest, national parks, nature reserves and other protected areas, such as those of special environmental, scientific, historical, cultural or spiritual interest.
Less strictly protected forests	COST E4	Protected forests according to some national protection category
Less strictly protected forests as in the IUCN protection categories III-VI. See further definitions at the Main Report or …	TBFRA-2000	▪ Natural monument (IUCN III) managed mainly for conservation of specific natural features ▪ Habitat/Species management area (IUCN IV) managed mainly for conservation through management intervention ▪ Protected landscape/ seascape (IUCN V) managed mainly for landscape/seascape conservation and recreation ▪ Managed resource protection area (IUCN VI) managed mainly for sustainable use of natural ecosystems
Strict forest reserves	COST E4	Protected for biodiversity. Areas in which no silvicultural operations or any other human impacts are allowed. In some cases the term strict may, however, contain areas where activities such as hunting, scientific research, ecotourism, control of unwanted species or amelioration of anthropogenic disturbances are carried out.
Strictly protected forests as in the IUCN protection categories I-II	TBFRA-2000	▪ Strict nature reserve/wilderness area (IUCN I). Protected area mainly for science and wilderness protection. These areas possess some outstanding ecosystems, features and/or species of flora and fauna of national scientific importance, or they are representative of particular natural areas. They often contain fragile ecosystems or life forms, areas of important biological or ecological diversity, or areas of particular importance to the conservation of genetic resources. Public access is generally not permitted. Natural processes are allowed to take place in the absence of any direct human interference, tourism and recreation. Ecological processes may include acts that alter the ecological system or physiological features, such as naturally occurring fires, natural succession, insect or disease outbreaks, storms, earthquakes and the like, but necessarily excluding man-induced disturbances. ▪ National park (IUCN II). Protected area managed mainly for ecosystem protection and recreation. National parks are relatively large areas, which contain representative samples of major natural regions, features or scenery, where plant and animal species, geomorphological sites, and habitats are of special scientific, educational and recreational interest. The area is managed and developed so as to sustain recreation and educational activities on a controlled basis. The area and visitors' use are managed at a level which maintains the area in a natural or semi-natural state.

Forests not available for wood supply due to conservation reasons

In Europe, 85 % of the forests are available for wood supply according to the TBFRA-2000 Main Report. Large areas (more than 20 % of the forest area) outside the wood supply can be found in 11 countries (Malta, Cyprus, Iceland, Portugal, Bosnia and Herzegovina, Israel, Liechtenstein, Italy, Norway, Spain and Sweden). On average 9.9 % of the forests are not available for wood supply due to conservation or protection reasons (Figure 4). In the EU-countries this figure is a bit lower, 6.5 %[8]. Generally, the number and size of protected areas has grown considerably since World War II, especially in Western Europe (UNEP 1999).

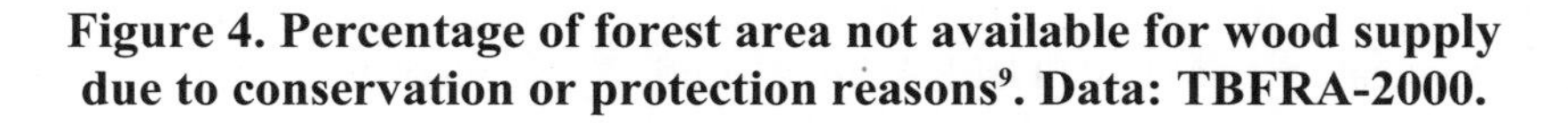

Figure 4. Percentage of forest area not available for wood supply due to conservation or protection reasons[9]. Data: TBFRA-2000.

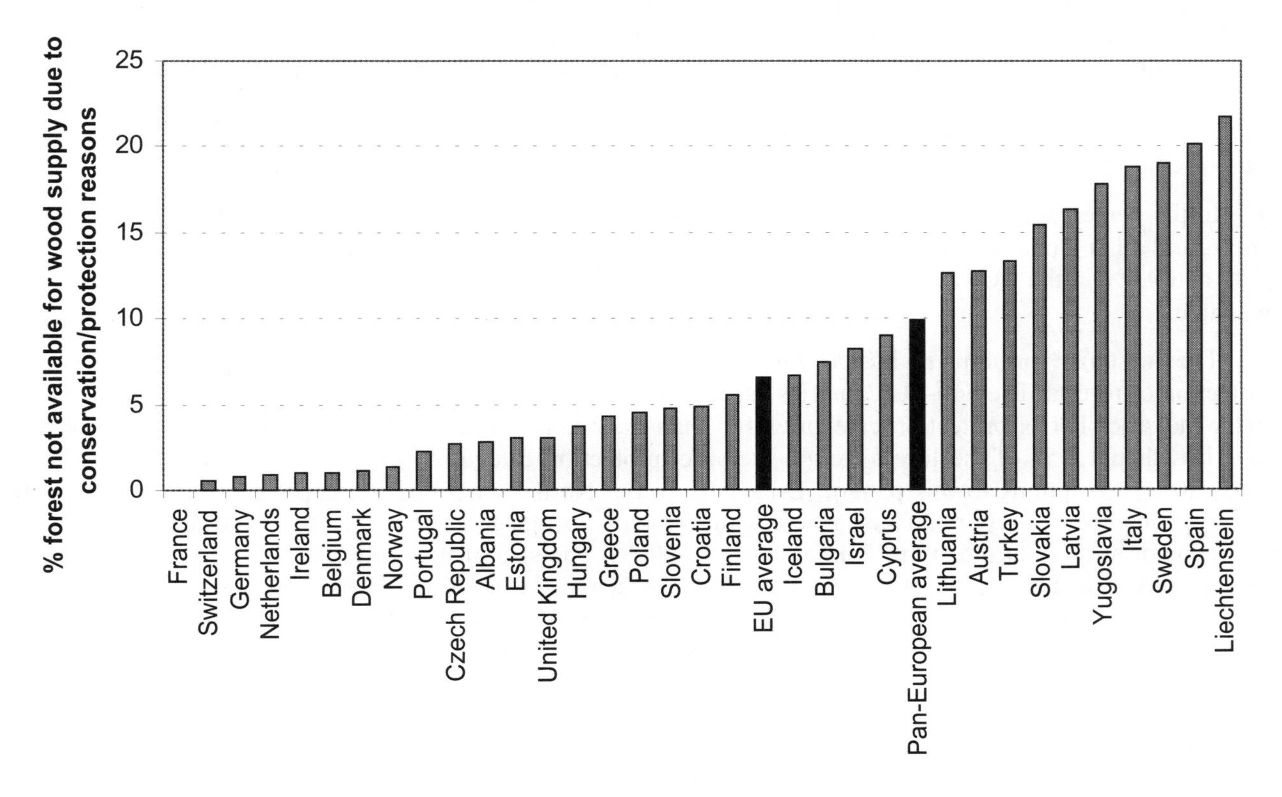

[8] Luxembourg is excluded from the figure due to missing data.

[9] The information is missing form Luxembourg, Bosnia and Herzegovina, Romania and the FYR of Macedonia. In Malta the total forest area is not available for wood supply due to the conservation/protection reasons.

Estimates for the amount of all protected forests and strictly protected forests

Table 6 summarises the information on protected forest and other woodland collected both in the TBFRA and in the COST Action E4 on 'Forest Reserves Research Network'[10]. The results support the assumption that meaningful comparisons of the protected forests within Europe require further clarification and analysis. Part of the differences can be explained through the different conception of protected forests. The COST Action E4 did not include forests protected for other purposes such as soil, ground water or avalanche protection. Instead, the Action tried to estimate the extent of forests protected primarily for biodiversity conservation. This, however, does not explain the whole extent of differences.

If all protected categories are considered, 23.8% of forests and other wooded land are protected according to the IUCN protection categories 1-6 (TBFRA-2000) and 7.3% according to the national definitions as collected in the COST Action E4. These results have been reported for a different number of countries, i.e. 34 and 26 countries, respectively. The difference in the number of observed countries, however, does not explain the large differences in the results. If only those 24 countries are analysed for which the data are available from both sources, the average percentage of protected areas from forest and other wooded land is 21.6% as reported according to the TBFRA-2000 (equals to 32 million ha) and 7.7% as reported according to the COST Action E4 (equals to 11 million ha).

In the percentage of the strict protected areas there is also a difference. In the TBFRA-2000 3.6% of the forests and other wooded land have been classified in the strictly protected IUCN I-II categories in the TBFRA-2000 and 0.9% in the COST Action E4. These figures are based on those 21 countries for which both data are available.

The Western, Northern and Eastern European countries seem to have a somewhat different approach to the protection of forests. The non-EU countries especially in Eastern Europe have a larger proportion of forest area under the strict IUCN protection categories I-II than the EU-countries (Figure 5). Most of the EU-countries have less than 1.8 % of the forest area in the strict protection categories. Only Finland, (Sweden), Italy and Portugal are exceptions to this pattern. Especially in the Nordic countries forest protection has targeted the preservation of old forest remnants, whereas in Western Europe forests are mainly protected as part of the landscape or as a cultural feature (Parviainen et al. 2000). The size of the protected forest reserves also varies considerably– so that the average size of a strict forest reserve is 30 – 50 ha in Switzerland, Poland and Germany, 123 ha in the UK and 375 ha in Spain. The largest single reserves of up to 70,000 ha are located in Finland and Sweden – and the smallest ones in Europe may cover only 0.5 ha (Parviainen et al. 2000).

[10] The area applied for the amount of forests and other wooded land in a country has been extracted from the TBFRA-2000 Main Report and the same figure has been used throughout.

Figure 5. Percentage of forest area under strict protection in different European countries. Data: TBFRA-2000.

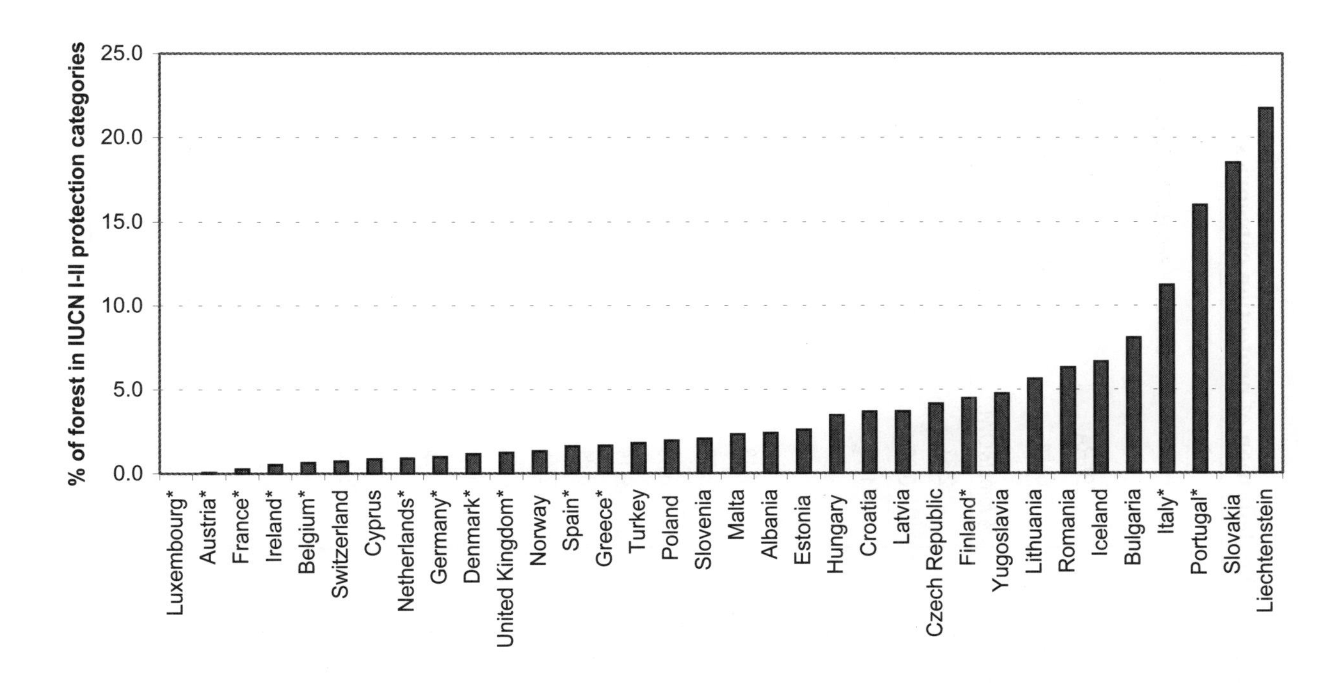

Table 6. Comparative data on forests in different protection categories in Europe*. For the definition of protection categories see Table 5

Protection type	FORESTS NOT AVAILABLE FOR WOOD SUPPLY		ALL PROTECTED FOREST AREAS				STRICTLY PROTECTED FORESTS			
Data source:	TBFRA		TBFRA		COST E4		TBFRA	COST E4		
Protection category:	Conservation or protection reasons		IUCN 1-6		National protection categories		IUCN 1-2	Strictly protected for biodiversity		
	Area 1000 ha	%forest**	Area 1000 ha	%forest&OWL*	Area 1000 ha	% forest&OWL**	% forest&OWL*	%forest&OWL**	Area 1000 ha	No of reserves
Albania	29	3	142	14	164	15.9	2.37	1.41	15	4
Austria	488	13	775	20	49	1.2	0.05	0.21	8	191
Belgium	7	1	204	30	5	0.7	1.47	0.19	1	36
Bosnia and Herzegovina	-	-	-	-	26	0.9	-	-	-	-
Bulgaria	265	7	1,354	35	335	8.6	7.43	-	-	-
Croatia	85	5	489	23	181	8.6	3.09	0.14	3	32
Cyprus	11	9	280	100	-	-	0.36	-	-	-
Czech Republic	71	3	646	25	175	6.7	4.14	0.95	25	103
Denmark	5	1	102	19	92	17.1	0.94	1.13	6	ca. 300
Estonia	61	3	191	9	-	-	2.45	-	-	-
Finland	1,208	6	2,527	11	2,440	10.7	4.46	6.72	1,530	311
France	0	0	3,346	20	180	1.1	0.67	0.08	14	30
Germany	83	1	7,205	67	400	3.7	0.98	0.23	25	679
Greece	142	4	1,220	19	952	14.6	1.06	2.18	142	39
Hungary	68	4	362	20	370	20.5	3.45	0.20	4	63
Iceland	2	7	14	11	-	-	3.85	-	-	-
Ireland	6	1	6	1	6	1.0	0.51	0.97	6	34
Israel	10	8	-	-	-	-	-	-	-	-
Italy	1,855	19	2,040	19	560	5.2	11.21	0.57	62	119
Latvia	471	16	488	16	-	-	3.67	-	-	-
Liechtenstein	1.5	22	2	27	-	-	24.32	-	-	-

Protection type	Forests not available for wood supply		All protected forest areas				Strictly protected forests			
Data source:	TBFRA		TBFRA		COST E4		TBFRA	COST E4		
Protection category:	Conservation or protection reasons		IUCN 1-6		National protection categories		IUCN 1-2	Strictly protected for biodiversity		
	Area 1000 ha	%forest**	Area 1000 ha	%forest&OWL*	Area 1000 ha	% forest&OWL**	% forest&OWL*	%forest&OWL**	Area 1000 ha	No of reserves
Lithuania	249	13	305	15	-	-	5.60	-	-	-
Luxembourg	0	0	1	1	-	-	0.00	-	-	-
Malta	0.347	100	0	10	-	-	2.31	-	-	-
Netherlands	3	1	80	24	19	5.5	0.88	0.89	3	60
Norway	114	1	4,555	38	200	1.7	1.67	1.23	148	160
Poland	398	4	1,405	16	183	2.0	-	0.04	4	106
Portugal	76	2	587	17	560	16.2	15.66	0.08	3	6
Romania	-	-	469	7	527	7.9	5.95	-	-	55
Slovakia	310	15	832	41	270	13.3	18.43	0.76	15	76(19)
Slovenia	52	5	84	7	71	6.1	2.09	0.89	10	186
Spain	2,727	20	3,211	12	3,000	11.5	0.83	0.13	33	87
Sweden	5,180	19	-	-	832	2.8	-	1.90	576	849
Switzerland	7	1	44	4	14	1.1	0.73	0.08	1	39
The FYR of Macedonia	-	-	-	-	-	-	-	-	-	-
Turkey	1,319	13	330	2	-	-	1.53	-	-	-
the United Kingdom	75	3	794	32	129	5.2	1.21	0.40	10	81
Yugoslavia	515	18	3,480	100	-	-	4.77	-	-	-

** Forest protection categories under review are 1) forests not available for wood supply due to conservation reasons (TBFRA-2000 data only), 2) all kinds of protected forests and 3) strictly protected forests (for the latter two both TBFRA-2000 data and data from the COST Action E4 on "Forest Reserves Research Network" (Parviainen et al. 2000) are provided). In the COST Action E4 principally only forests protected for biodiversity have been considered.*

*** The data from the TBFRA-2000 have been used for area of forest and OWL in a country.*

4.3 Species Mixtures

Mixture of species and crown layers increase the horizontal and vertical structure of a forest stand. An increase in the structural variety is currently seen as a way to increase the ecological and mechanical stability against large storm damages and the effects of climate change in Europe. There is still a gap in fully understanding the relationship between forest structure, biodiversity and ecological stability (e.g. Pretzsch 1997 and 1999a). It is, nevertheless, assumed that mixed species and crown layer structure can support a larger variety of different species compared with a single species one-layered stand. Such mixtures may improve on one hand the protection against biotic damage caused for instance by insects and on the other hand the dissemination of risk related to abiotic damages caused for instance by emissions, wind and snow. There may be a certain productivity advantage in growing and managing mixed instead of single species stands. Mixed forests are gaining popularity also due to their biodiversity potential and amenity value, and because they are thought to approximate natural forests and natural ecosystem functions more closely than single-species forests in many parts of Europe (Bartelink and Olsthoorn 1999).

Natural species mixture and geographic location

Countries' species composition and distribution reflect their geographic location in terms of latitude and altitude at which the forests grow. Tree-species composition and dynamics in natural forests are primarily a result of climate and secondarily of soil parent material, controlled by water resources (e.g. Kuusela 1994). Even in managed forests the natural forces within the ecosystem affect the stand to some extent – and increasingly if the stand is left for free development. Therefore species mixture alone does not guarantee a higher stability or naturalness of a forest. Instead, the natural composition of the forests in a particular region and location should be considered when defining silvicultural strategies.

In high latitudes and altitudes single-species forests naturally dominate. Coniferous species are prevalent in Europe especially in the Nordic countries, in countries with notable mountainous areas (Austria, Germany, Switzerland, Poland), in certain habitats such as peatlands or poor soils where other species do not grow, and in the British Isles where conifer plantations enhance the proportion of conifers. Generally, conifers are well adapted to cold climates due to their ability to begin photosynthesis at lower temperatures than broadleaved trees, and due to the immediate photosynthesis after the temperature is high enough, i.e. there is no need to grow leaves first (e.g. Kuusela 1994).

As the temperature increases broadleaved trees become more competitive (Figure 6). Yugoslavia has the highest percentage of broadleaved species with 87%, followed by Croatia with 82%, Hungary with 78%, Italy with 72%, Romania with 70%, Bulgaria with 67% and France with 64%. Mixed forests are naturally most frequent in Europe in broadleaved deciduous and in mixed evergreen forest zones (Kuusela 1994). The former are located between the latitudes 40° N and 60° N and the latter comprises mainly the Mediterranean countries.

During the last two centuries the forest management and forest establishment in Europe often favoured single-species monocultures consisting of one crown layer. Currently, the area of mixed forests is most likely increasing due to deliberate transformation of man-made monocultural plantations into mixed species forests and due to spontaneous change when natural processes such as natural regeneration are allowed to take place (e.g. Bartelink and Olsthoorn 1999). The success of natural processes in yielding species mixtures depends on the location. In some regions such as in boreal forest zone mixed forests are mainly associated with the early phases of the natural tree species succession (Leikola 1999). In Central European monocultures natural regeneration would commonly yield a new stand with a single species so that careful silvicultural measures are needed in order to transform plantations into naturally regenerative mixed forests on appropriate sites.

Figure 6. Estimate of the spatial distribution of deciduous forests in Europe*

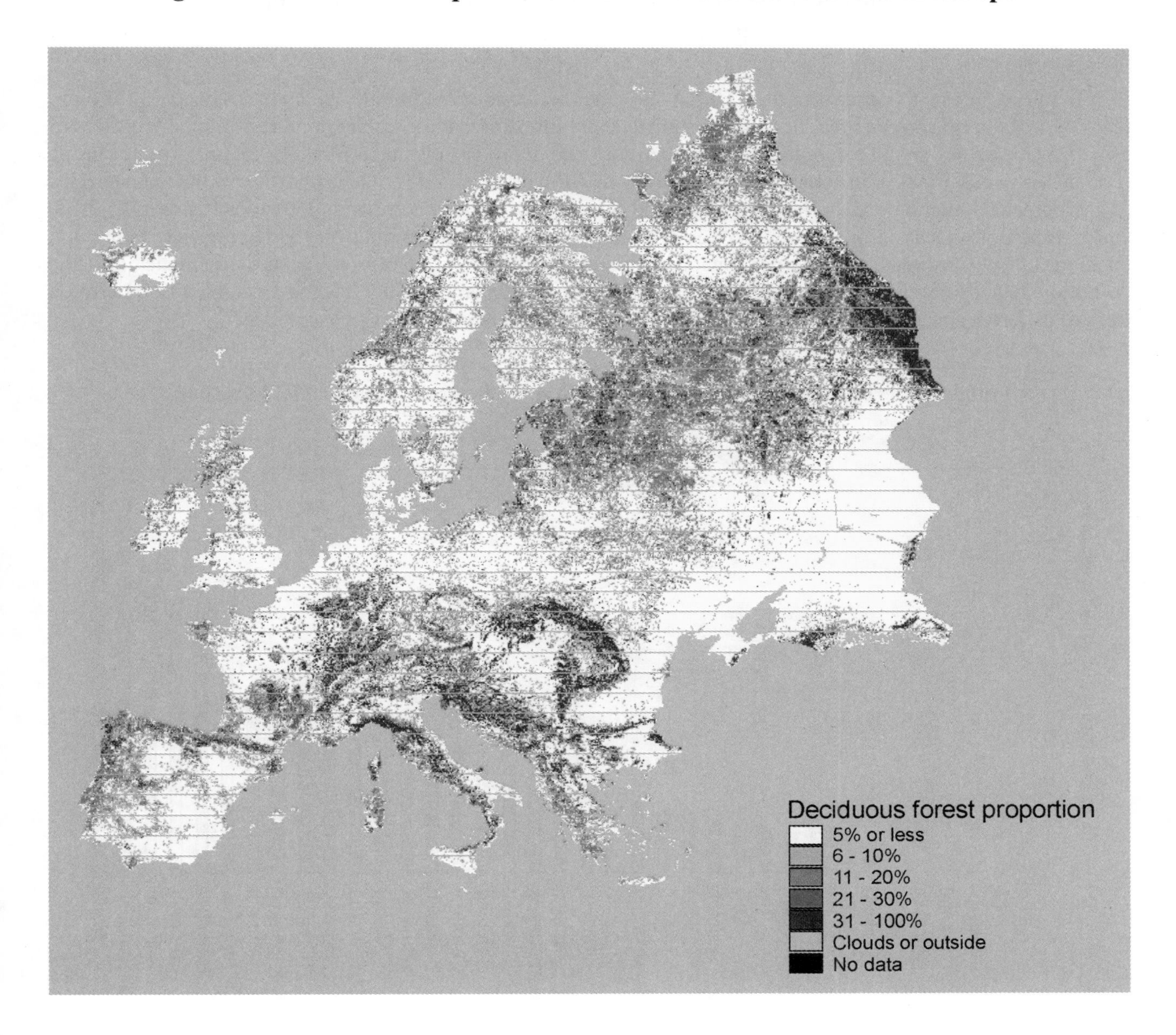

** The spatial map is an intermediate product of the JRC – European Commission financed on-going project "Forest and tree groupings data base of the EU 15 and pan-European area derived from NOAA-AVHRR data". The work is carried out by University of Joensuu, Technical Research Center Finland (VTT) and European Forest Institute. The map is based on the interpretation of 1 km resolution AVHRR data, and it will be refined further to match with the regional forestry statistics in Europe (Päivinen et al. 2000).*

Some estimates and definitions for the amount of mixed species forests in Europe

The current amount of mixed forests is about 14% of forest area in Europe, and somewhat lower if the EU-countries are investigated alone (8.5%) (TBFRA-2000). The latter corresponds to almost 15 million ha. In ten countries in Europe the percentage of mixed forests is larger than 20% of the total forest area (Figure 6). These figures are somewhat lower than the ones reported by Bartelink and Olsthoorn (1999). They estimate that there are roughly 30 to 45 million ha of mixed forests in Europe. The reason for the difference is the different definitions used for mixed forests, and possibly also a difference in the countries, which have been included in the analysis.

In mixed forests as defined in the TBFRA-2000 neither conifer not broadleaved species account for more than 75% of the tree crown area. In national statistics and forest inventory results mixtures of either conifers or broadleaved species are also counted. And furthermore, the definition of mixed forest is not uniform in Europe (Päivinen and Köhl 1997). In some countries there is no official threshold for the proportion of the main species (e.g. Denmark, Finland, Greece, Ireland, Spain and Switzerland), and if there is one it varies between 20-30%, and is based on the basal area, crown cover or volume proportion, depending on the country. Bartelink and Olsthoorn (1999) estimate that the percentage of mixed forests should be revised upwards dramatically if the definition "stands composed of different tree species, mixed on a small scale, leading to competition between trees of different species as a main factor influencing growth and management" was applied.

Figure 7. Percentage of mixed forests as total forest area. [11] Data: TBFRA-2000.

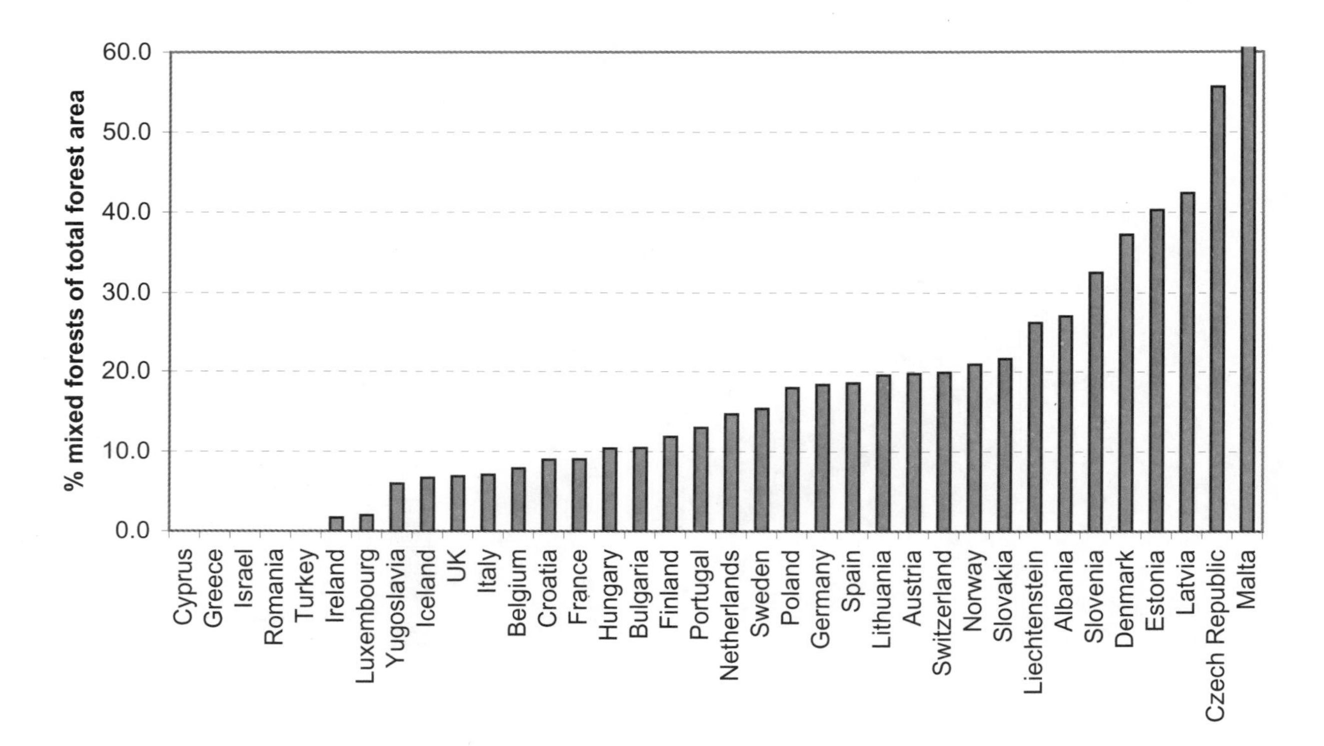

[11] In some countries such as Denmark, Turkey and Greece the inventories have distinguished between conifers and broadleaves only so that the information related to mixed forests need to be interpreted with care.

4.4 Age Structure

Forests at different ages provide diverse habitats for various species ranging from insects dependent on newly burned ground and pioneer tree species to climax tree species and species, which need decaying wood and old trunks. Therefore, a mixture of age classes is needed to support a wide range of species compositions and development processes. Age structure provides also information about the future production possibilities, indicates if the usage of timber resources is on a sustainable basis and allows the assessment of the proportion of old growth forests. In the TBFRA-2000 assessment information on the coppice forests and high forests has been collected. High forests have further been divided into even-aged and uneven-aged forests.

Coppice forests

Coppice stands are featured by a rather short rotation time (often 20 – 30 years), rapid growth in the first 3-5 years when the coppice sprouts utilise the resources from the stool, and high social differentiation within a stand. Coppice is an old practice in Europe and in some area belongs to the cultural landscape. Coppice structures support species, which are associated with the early development stages of forests. In coppice management mature trees are cut, and a new stand is regenerated from the re-growth of coppice stools. Coppice with standards means that two different rotation times are applied in a stand; one (shorter) for the coppice sprouts and the other for the scattered trees, so called standards, which originate either from seeding or from sprouting.

Currently, considerable coppice areas can be found especially in Southern and South-Eastern Europe (Figure 8). In total coppices cover about 21 million ha or 16% of forests available for wood supply. The coppice area in France alone is almost 7 million ha, followed by Italy (almost 3.5 million ha) and Greece (over 2 million ha).

Figure 8. Countries in Europe with a considerable coppice area. Data: TBFRA-2000.

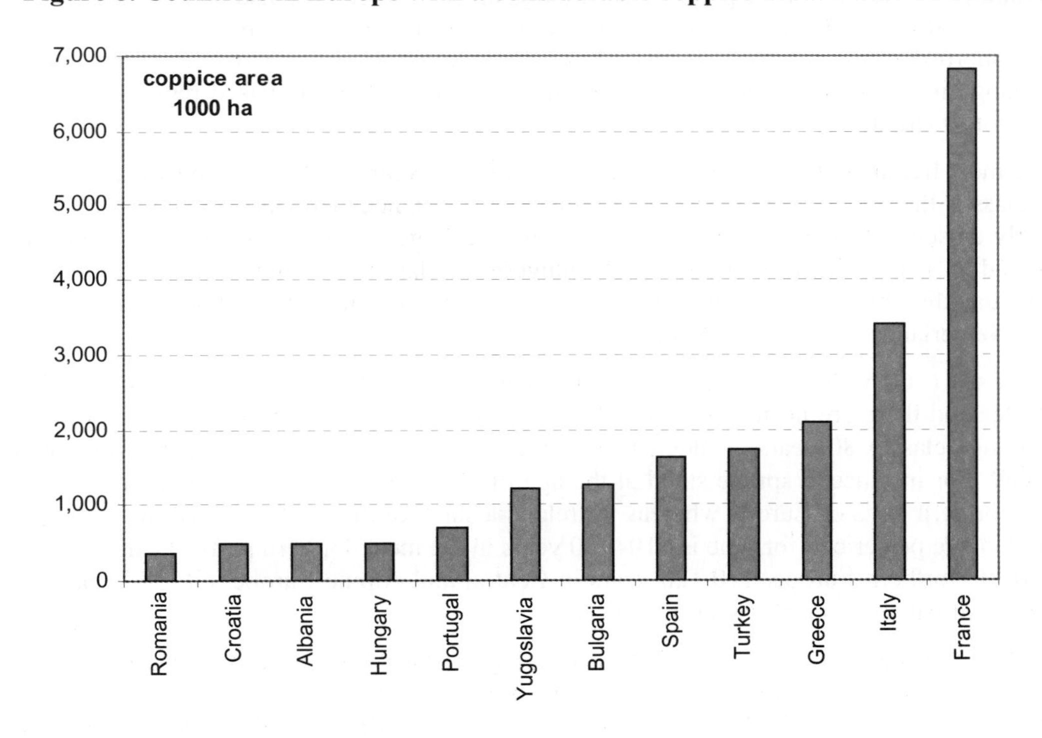

Even-aged high forests

General pattern

The average age class structure of even-aged high forests in Europe is presented in Figure 9. The largest forest areas are found in age classes 20-80 years, and especially the proportion of forests in age classes 20 – 40 years and 40 – 60 years is emphasized. This peak is likely to have its origin in the large afforestations, which were carried out in Central Europe after the Second World War.

Old growth forests are important as special and possibly endangered ecosystems. There are almost 13 million ha of high forests older than 100 years of age in Europe, according to the TBFRA-2000. There are currently little spatial data on the location of these old forests. Based on the national information, however, it becomes clear that the old age classes are often located in areas, which are difficult to access (i.e. mountain regions or remote areas in Northern Europe) or in which the growth rate is so low that it takes a long time to yield merchantable timber dimensions.

Different age-class distributions

The overall age-class distribution encloses a large variety of age structures in different countries. In order to compare the age-class structure of different countries the age–class distributions have been transformed into a relative form, i.e. into proportions of the total area of high forest available for wood supply in a country. Figure 10 illustrates some typical patterns found in Europe. Firstly, countries with a distinct peak in the age class distribution and countries with an even age-class distribution (i.e. no one class possesses more than 1/5 of the total area) can be identified. An even distribution is found in Switzerland, Luxembourg, Czech Republic, Finland and Norway. Sweden and Albania are also very close to such an even distribution.

Secondly, the peak in the age-class distribution is in different age classes in different countries. Ireland is an example of a country where the percentage of the youngest age class is relatively high (20% of the forests in age class ten years or younger). Other countries with a similar pattern are Portugal, Israel and Iceland. In Austria (17%) and Denmark (18%) the percentage of the youngest age class is also considerable. These figures imply the increased afforestation during the last ten years, not necessarily increased harvesting activity. In Germany, the proportion of the youngest age class has been reported to be close to zero. This is likely to be an effect of the change in silvilcultural practices, which increasingly favour gap-based regeneration for uneven-aged and mixed forest structures.

In Italy the most frequent class is at the young age of 11 – 20 years. Most commonly (in 17 countries) the most frequent class is the one of 20-40 years, as in France. For instance Germany, Finland and Sweden follow this pattern fairly closely. Only the proportion of the two oldest age classes is in Germany somewhat smaller than in France and in Sweden the proportion of the youngest age class is a bit higher than in France. At the age of 40-60 years the distribution peaks in Latvia, Lithuania and Estonia as well as in Croatia, Slovakia, Yugoslavia, the Netherlands, Denmark and in Iceland.

Old age classes (>60 years) are relatively frequent in the Czech Republic, Switzerland, Slovenia and Luxemburg. In Iceland there are no forests in the age class 60 years or older, and in Israel and Portugal there are no forests in age classes 80 years or older. Due to the different ecological systems the rotation ages used vary considerably. For instance, a spruce stand at the age of 60 years is considered to be in the middle of the rotation age in Northern parts of Europe whereas in Ireland a stand of the same age is at the end of its rotation cycle. The rotation age prescribed for pine is 110- 180 years in the most northern parts of Europe, 80-120 years in southern part of Northern Europe, 110-120 years in Central and Alpine Europe, 60 – 70 years in Hungary and about 50 years in Atlantic Europe (Kuusela 1994).

Thus, a lack of old age classes from the country data may be due to the generally lower rotation age, due to intensive harvesting in the past or due to increased afforestation and regeneration in the past – or due to a combination of all these factors. The lack of old age classes does not necessarily imply that there are no old forests left in the country – it only indicates that old forests are not prevalent within high forests available for wood supply. Old forest structures may be found from uneven-aged forests, from forests not available for wood supply either due to conservation or economic reasons or from forests in protected areas.

Figure 9. Average age-class distribution of even-aged high forest available for wood supply in Europe. Data: TBFRA-2000.

Figure 10. Examples of different age-class distributions found in European countries[12]. Data: TBFRA-2000.

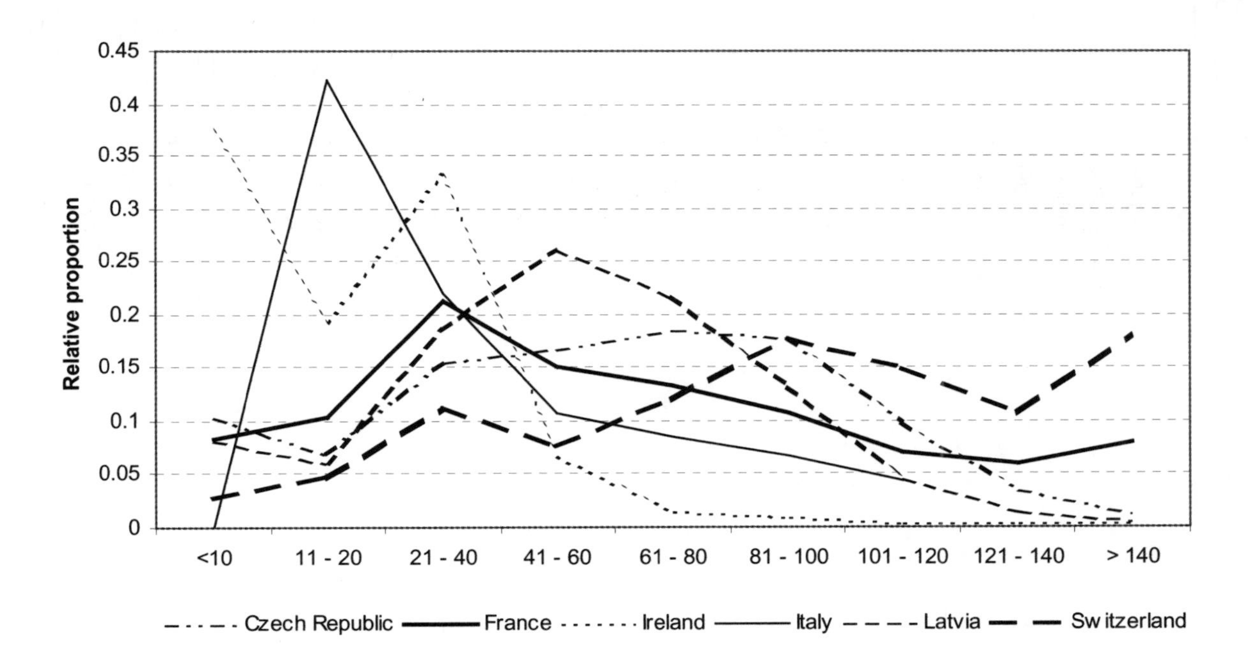

[12] The relative proportions of each class have been calculated from the even-aged high forest available for wood supply in each country.

Uneven-aged high forests

The prevalence of uneven-aged forests indicates structural variation within the forest. Several age classes, crown layers and development stages commonly occur at the same time, i.e. a stand is composed of trees of different sizes and ages. Uneven-aged forests may or may not contain species mixtures. Uneven-aged forests comprise 16% of the high forests available for wood supply in Europe (Figure 11). The percentage of uneven-aged mixed forests is moderate – on average less than 4% of the high forest available for wood supply. Slovenia (23%), Portugal (22%), Liechtenstein (15%), Yugoslavia (13%), Croatia (12%) and Sweden (4%) lie above the average proportion.

Figure 11. Percentage of uneven-aged forests as a total of high forests available for wood supply. Data: TBFRA-2000.

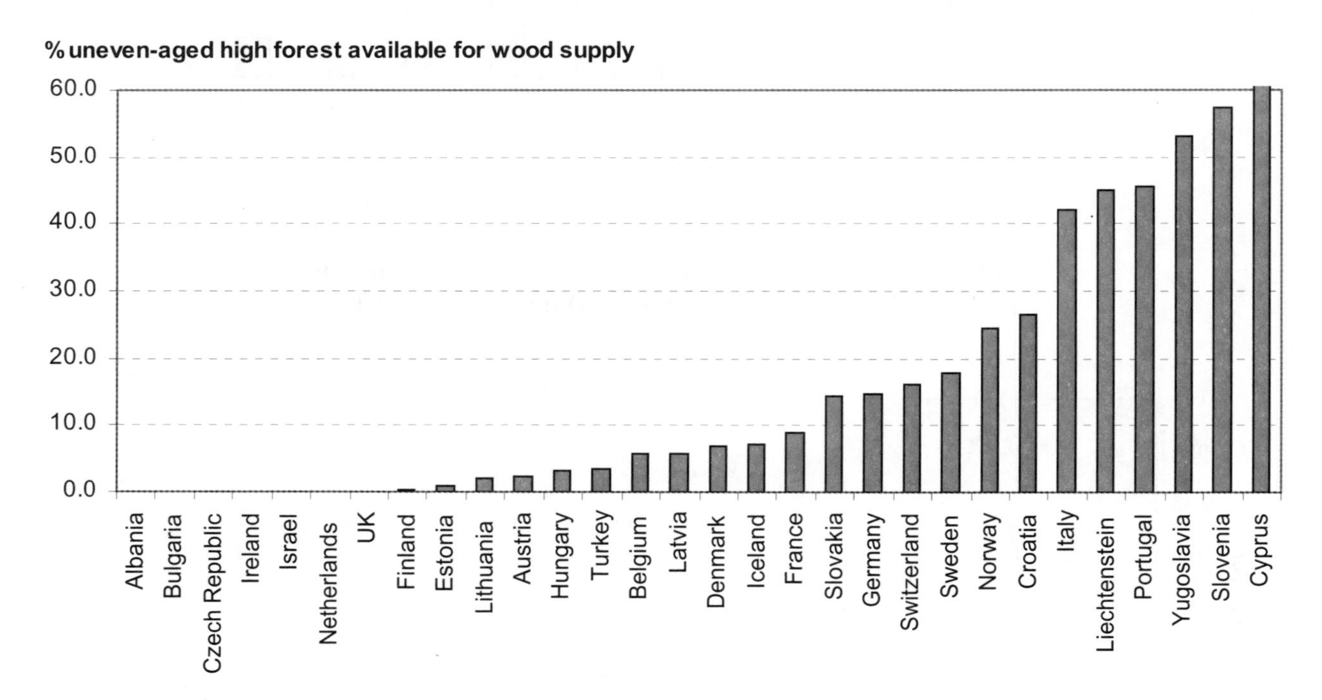

4.5 Regeneration, Extension and Natural Colonization

The forest area is slightly increasing in Europe, mainly due to the afforestation of former agricultural lands, due to the conversion of other wooded land into forests in the Mediterranean countries and due to afforestation programs to increase the wood supply through plantations in some countries. In total, the forest area has increased by about 500,000 ha annually, while the area of other wooded land has decreased by about 200,000 ha annually. The EU-countries comprise about 340,000 ha of the increase in forest area.

The creation of forest and other wooded land change not only the extent of forest area, but also the structure by replacing either former forest areas with young trees or replacing other land use forms with trees. The methods and species used in the creation of new forests determine the species composition, genetic variety and biodiversity not only at local but also at a national level, and affect the succession in the long term. "New" forests and other wooded land emerge through (TBFRA-2000 classification):

- *Regeneration* of land that has recently been forested (with natural regeneration, natural regeneration enhanced by planting, coppice sprouting or planting/seeding),
- *Extension* of forest on new areas (afforestation) or on former other wooded land (with natural colonization, natural conversion of other wooded land to forest or planting or seeding of non-forest or other wooded land),
- *Natural colonization* of non-forest land to other wooded land.

The proportions of the different forms of forest creation in different countries are shown in Figure 12. In most countries, regeneration plays a major role by replacing harvested areas as part of normal forest management practices. Extension and natural colonization are important especially in Iceland, Ireland, Israel, Malta, Turkey, the United Kingdom, Norway, Portugal and Estonia. The countries where over 50,000 ha of forest has been established annually are Sweden (206,000 ha), France (192,000 ha), Finland (188,000 ha), Italy (143,000 ha), Portugal (133,000 ha), Turkey (120,000 ha), Germany (77,000 ha), Poland (69,000 ha) and Austria (52,000 ha). These figures include regeneration, extension and natural colonization.

Figure 12. Regeneration, extension and natural colonization as percentage of total forest area created annually[13]. Data: TBFRA-2000.

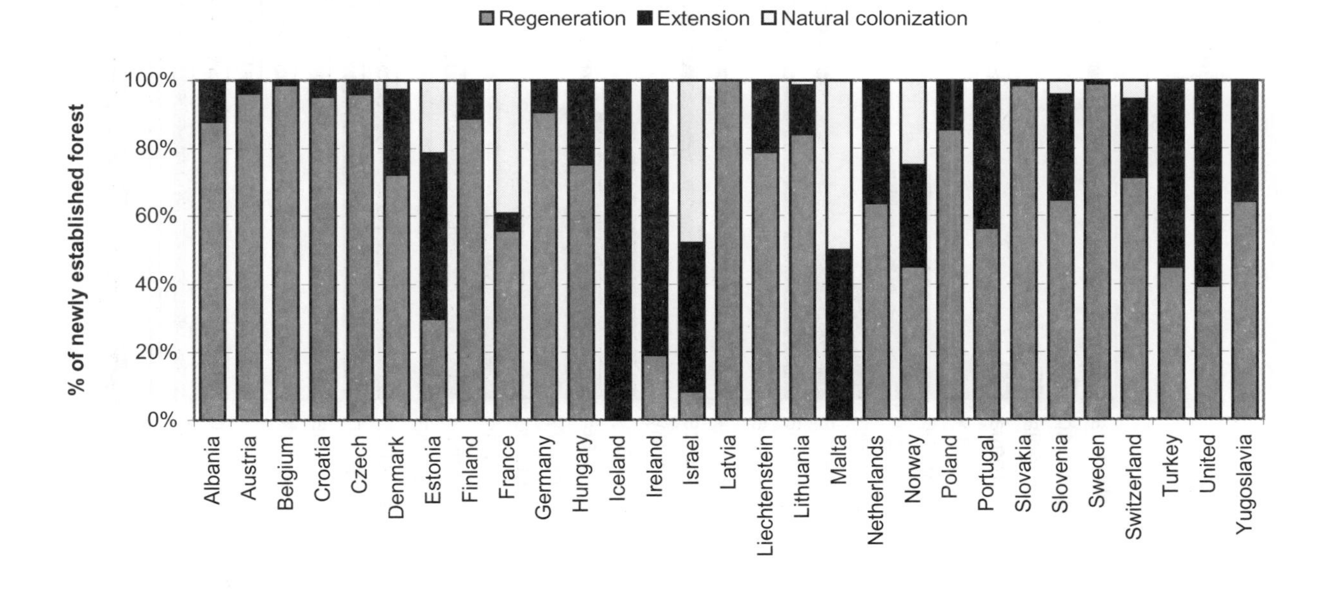

[13] Data on extension and natural colonization area are missing for Cyprus, Italy and Bulgaria. For Finland, Hungary, Iceland, Poland, Switzerland and the UK only figures for regeneration and extension are available.

Planting and seeding

Planting and seeding are generally seen as "less natural" and often in the context of biodiversity as "not so good" as natural regeneration. This interpretation, however, is not so straightforward. For instance, careful usage of planting or seeding often under the shade of mature trees can be applied to transform monocultural plantations into mixed species forests – or to maintain the monocultures. Therefore, the local application of planting and seeding should be studied prior to assessing the methods as "good" or "bad" in terms of promoting biodiversity.

The selection of seeds and planting material has a major impact on the biodiversity of the future forests. Generally, there is a trend towards controlled seed origin so that regionally adapted seeds with high quality, quantity and resistance, especially related to timber growth, are used. This, on one hand, contributes to high growth levels and resistance but, on the other hand, does not necessarily contribute to a large diversity of the genetic material, especially if fairly uniform parent material is used.

Planting and seeding play a clear role in forest regeneration and extension in Europe, but they are used to a varying degree in different countries (Figure 13). In regeneration more than one half of the area is planted or seeded in 17 countries, especially in Northern and Western Europe. In Southern and South-Eastern Europe natural regeneration and coppice sprouting tend to play a larger role. Forest extension is being taken care of almost uniformly by planting and seeding in 16 countries. Data are missing for Spain, Greece and Romania. However, figures on the change of the area of forests and other wooded land are available. In Spain the forest area has increased by 86,000 ha, whereas the area of other wooded land has decreased by about 68,000 ha. In Greece these figures are 30,00 ha and 29,000 ha, respectively. In Romania the forest area has increased by about 15,000 ha.

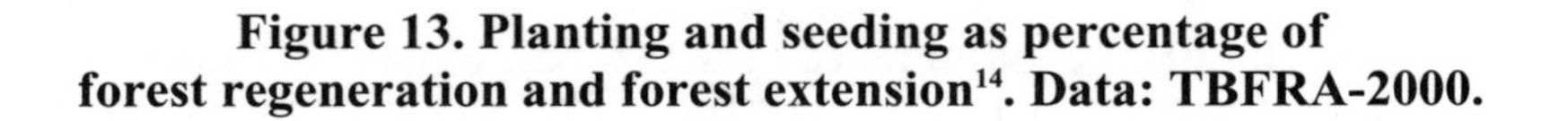

Figure 13. Planting and seeding as percentage of forest regeneration and forest extension[14]. Data: TBFRA-2000.

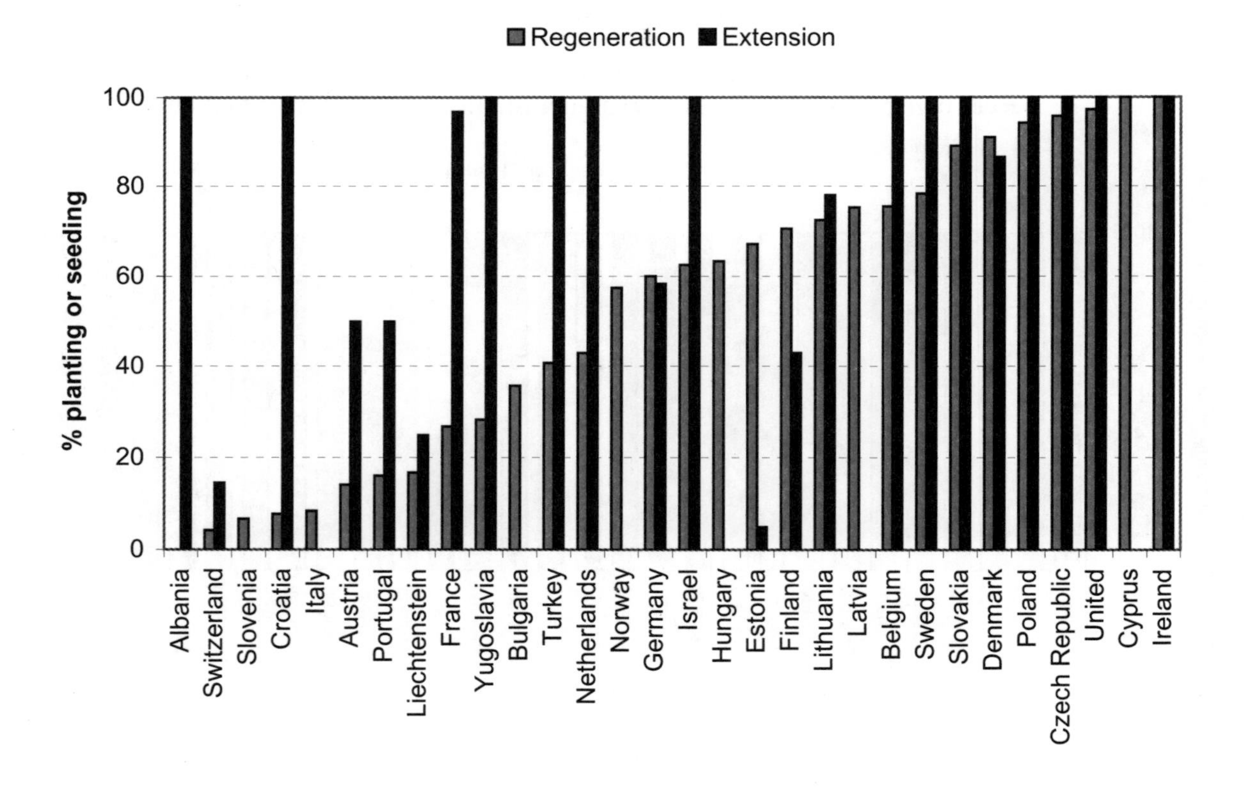

[14] Forest extension includes planting and seeding of both non-forest and other wooded land.

4.6 Summary of Structural Factors

European countries show structural variety in different ways. To illustrate this, figures 14 and 15 summarise different aspects of structural variety in one graphic. The figures underline the importance of multi-criteria approach in the evaluation of the structural variety and biodiversity. A single country may have on the one hand a large proportion of mixed forests, but on the other hand mainly even-aged forests (e.g. Latvia and Slovakia). In another country these factors may be the opposite (e.g. Portugal and Italy). A country may have a relatively large percentage of strictly protected forests, whereas the total area of forests not available for wood supply due to conservation reasons may be moderate compared with another country.

The EU-15 countries on average have a little more plantation forests and use more introduced species and planting and seeding in establishing new forest compared with the whole of Europe (Fig. 16, upper diagram). In the whole European area, there is a larger proportion of mixed forests, forests not available for wood supply due to conservation reasons and strictly protected forests, than found on average in the EU-15 countries (lower diagram). In the proportion of coppice there is no clear difference. The EU-15 average for the proportion of uneven-aged forests is clearly higher than the average in whole Europe.

This kind of variety even at such a low level of resolution like the national level emphasizes the fact that one or only a few indicators rarely can suffice and describe the true variety in forest structure in Europe, and that the attempts to assign one comparable value for the forest biodiversity are very difficult. The large variety is, on one hand, due to the ecological factors and differences in climate, soil and water resources and, on the other hand, due to current and past anthropogenic changes. For instance, different political approaches have contributed to varying forest ownership structures, to different proportions of conserved forest area and to the allocation of conserved area into categories ranging form strict to less strict protection status, for instance.

Figure 14. Some structural key factors presented for selected European countries[15]

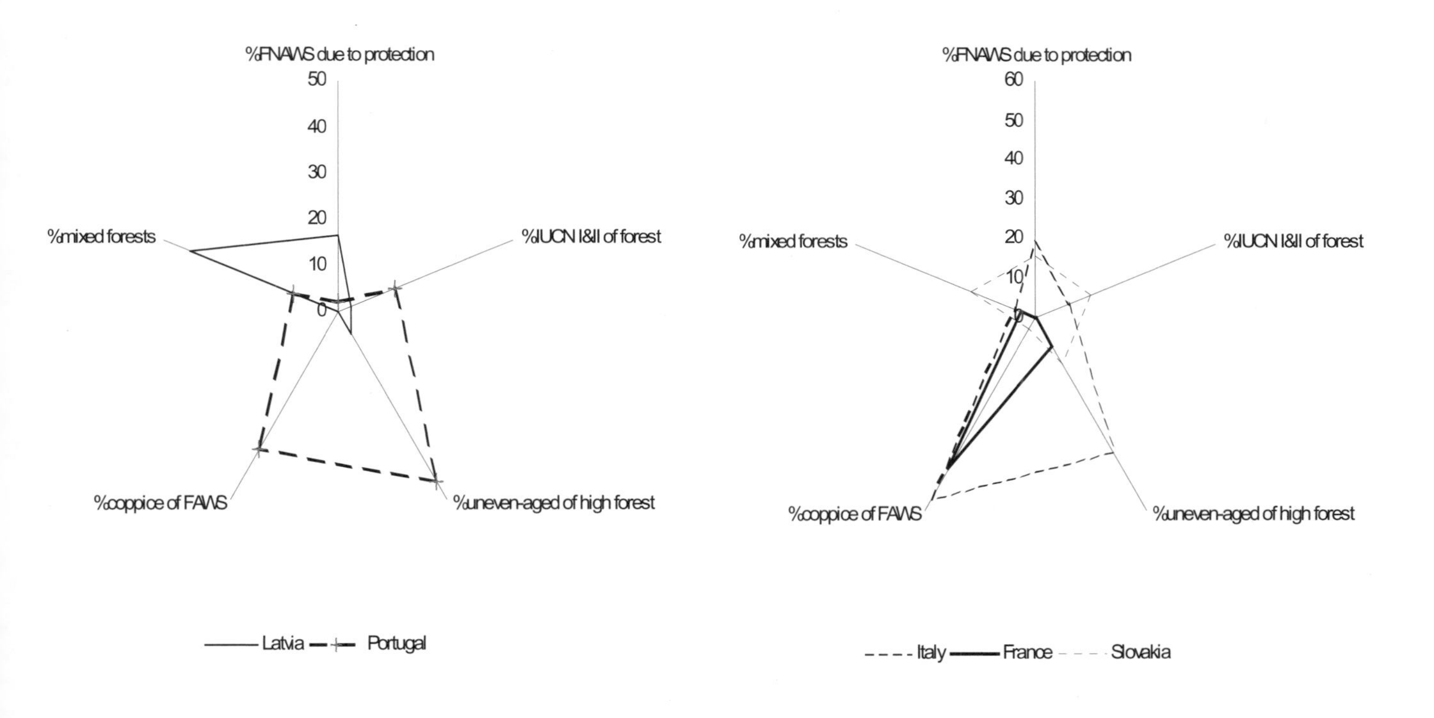

[15] The structural factors are percentages of forests not available for wood supply (FNAWS) due to conservation reasons, of forests in the IUCN protection categories I-II, of uneven-aged forests of the high forests, of coppice from the forests available for wood supply (FAWS) and the percentage of mixed forests. Data: TBFRA-2000.

Figure 15. European and EU-averages for selected structural key factors of biodiversity

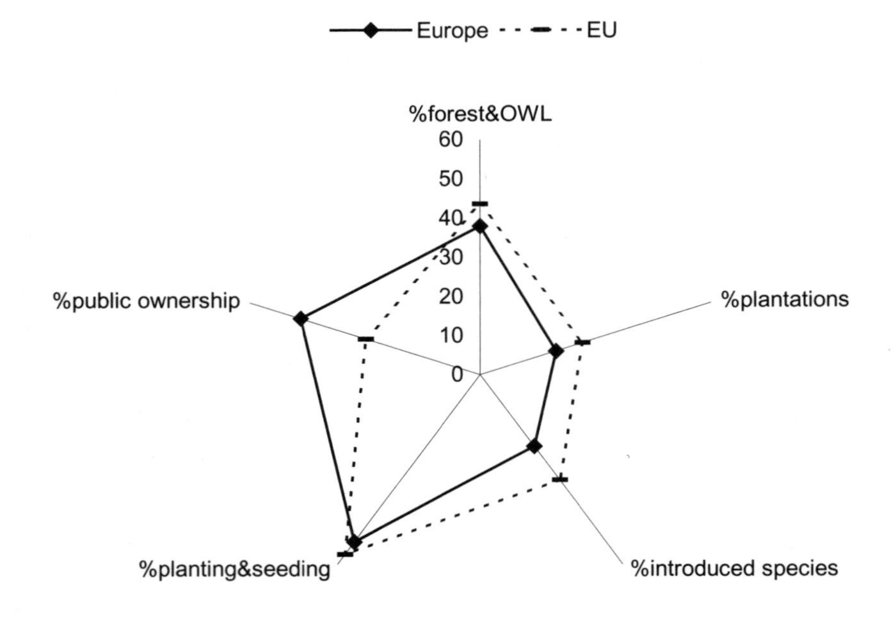

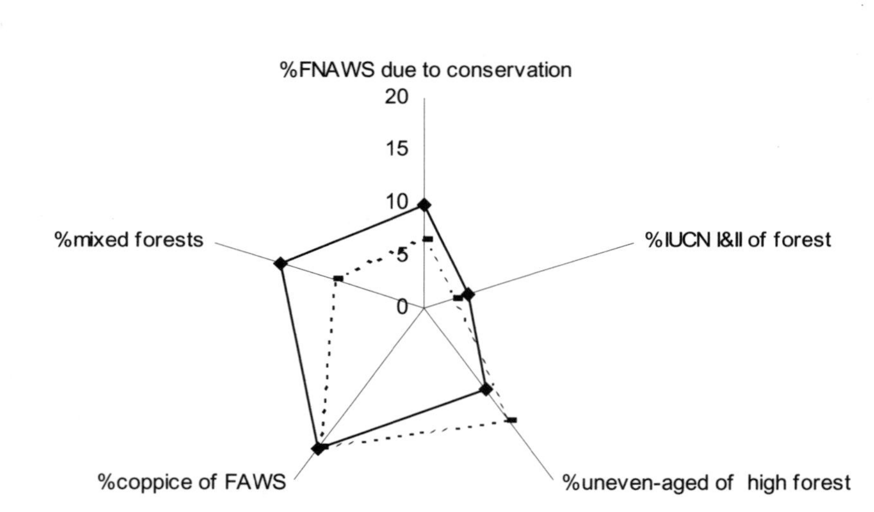

Note:

All the five axes (%) in each diagram are on the same scale.

Variables of the upper diagram: % of forests and OWL from the total land area (%forest&OWL), forests and OWL in public ownership (%public ownership), % of plantations from the total forest area, % of introduced species used in forest establishment (in regeneration, extension and natural colonization) and % of planting and seeding used in regeneration.

Variables of the lower diagram: % of FNAWS due to conservation reasons, % of forests in the IUCN protection categories I-II, % of uneven-aged forests of the high forests, % of coppice from the forests available for wood supply (FAWS) and % of mixed forests. Data: TBFRA-2000.

5. Compositional Aspects of Forest Biodiversity in Europe

Composition as an element of biodiversity relates most commonly to the number of species[16]. However, besides the sole count of species number the understanding of the relationships between species and structures found in an area, and the functioning of the species in their ecological niche are increasingly important. All taxonomic groups are not necessarily "biodiverse" in the same habitats or at the same time - so that either a set of indicators or an umbrella species representing the requirements of a group of species is needed. Different taxonomic groups from vascular plants, birds and butterflies to carabid fauna have been proposed as bioindicators (e.g. Furness and Greenwood 1993, Thomas 1995).

The European contribution to the total number of species in the world is relatively moderate. Only some 2-6% of the world's species are present in Europe, varying according to the species group (EEA 1999). The proportion of species occurring only in Europe, however, is considerable for several species groups. 75% of amphibians, 58% of freshwater fish and 45% of reptile species present in Europe occur only in Europe. The same applies to about one third of the mammal, butterfly and vascular plant species. Only 6% of the breeding bird species are endemic to Europe, but Europe also forms an important seasonal home for a huge number of migratory species.

Since the species data for several countries are missing, it is not possible to conclude a European assessment for the total number of species or threatened species based on the TBFRA-2000 alone. According to the EEA (1999) about 12,000 vascular plant species are reported to be known in Europe. Of these 3,500 species occur only in Europe. In the majority of European countries there are less than 150 different tree species (TBFRA-2000). The total number of fern species range between 17 in Iceland and 114 in Portugal, the number of moss species from 20 in Malta and 234 in Turkey to over 1,000 species in France, Germany, Norway, Spain, Sweden, Switzerland and the United Kingdom (TBFRA-2000 Main Report). The largest numbers of endangered vascular species are in Israel (408), Slovakia (360), Austria (271), Yugoslavia (217), Switzerland (110) and Lithuania (110) (TBFRA-2000). Additionally, more than one fifth of the total number of vascular species is endangered in Belgium, Estonia, Latvia, The Netherlands and in Sweden. Less than 5% of the species have been classified as endangered in Albania, Cyprus, France, Germany and Portugal.

A total of 270 mammal species, 607 other vertebrate species, 514 breeding bird species and about 575 butterfly species are reported to be know in the European Union area (EEA 1999). Of these 44 mammal species, 111 other vertebrate species and 17 bird species have been classified as threatened species (EEA 1999). The majority of the species are vulnerable (113 species) and the number of critically endangered species amount to 26. Additionally about 12% (71 species) of the European butterfly species are classified as endangered (Swaay Van and Warren 1999).

---

[16] For instance the Pan-European indicators for sustainable forest management include "Changes in the number and percentage of threatened species in relation to the total number of forest species" (see Table 2).

5.1 Availability and Validation of the Species Number Data in the TBFRA-2000

Features of the TBFRA-2000 species data

In the TBFRA-2000 assessment the countries have provided species data on the total number of species and on the number of endangered species for the following flora and fauna categories: Trees, Other vascular plants, Ferns, Mosses, Lichens, Mammals, Birds, Other vertebrates and Butterflies and moths. Parallel to the total numbers also the numbers of endemic species have been reported.

The TBFRA-2000 assessment differs from, and complements in important aspects, the international species richness assessments. Even if the data are missing for some countries and not completely reliable for some others, they still allow an interesting scope of observation. Especially the following elements are of special interest:

- Information has been collected not only on the total number of different species or endangered species, but also on the number of forest occurring species. Most other international processes and data sources deliver only the total number of species, independent of the habitat. Thus, the TBFRA-2000 data should allow an investigation of the role of forests and other wooded land as a habitat for different species.
- Separate data have been collected on the number of different tree species. Trees form the basic structure of forests and their mixture supports further structural variety within a forest stand.
- An attempt has been made to collect data on the number of moss and lichen species.

Even if the number of species seems to be a simple indicator at the first glance, the reported numbers are not in all cases consistent and therefore comparisons should be carefully made. "The distribution of species has temporal as well as spatial dimensions, and species will only be detected if the observer is using the right method, at the right place, at the right time" (Vanclay 1998). All the species are not known and the recognition of the species is difficult. For instance, in Israel the current number of recorded wild vascular plant species is 2,780, and a further 110 species are assumed to be found in the future (Gabbay 1997). Furthermore, the figures reported for Europe depend on the definition of Europe – and this varies between the different international processes. National figures are less ambiguous even if some changes have taken place in the national borders in Eastern Europe during the last decade. In addition, species groupings and definitions applied may be different[17].

In the following, an attempt is made to validate, compare and complement the figures reported in the TBFRA-2000 by using figures reported in

- international processes (i.e. WCMC 1994 and EC 1995),
- recent reports on national biodiversity and
- other data sources.

Data from several sources, where available, have been sought after to verify the similarities and discrepancies, the reason for the discrepancies, and to assess the credibility of the reported figures. Since the dates, methods and definitions used in the monitoring have been different, there is little point in trying to identify trends based on the figures reported in different processes. The differences seem to originate mainly from different definitions or interpretations and from improved information, which has become available during the last decade due to the increased emphasis on the assessment of biodiversity.

---

[17] For instance the TBFRA-2000 distinguishes between mammals, birds, other vertebrates and butterfly species whereas for instance IUCN, WCMC, UNEP and EEA reporting further distinguishes other vertebrates into reptiles, amphibians and fishes. The total number of birds may refer to the total number of visiting or breeding species.

5.1.1 Endangered species

When is a species is classified as endangered? Prior to 1994 the IUCN applied threatened species categories endangered, vulnerable, rare, intermediate and insufficiently known taxa. The survival of endangered taxa "is unlikely if the causal factors continue operating. Included are taxa whose numbers have been reduced to a critical level or whose habitats have been so drastically reduced that they are deemed to be in immediate danger of extinction. Also included are taxa that may be extinct but have definitely been seen in the wild in the past 50 years". "Vulnerable" are taxa likely to move into the "endangered" category in the near future if the causal factors continue operating. The number of vulnerable species is for most species groups larger than the number of critically endangered or endangered species (see e.g. UNEP 1999, EC 1998).

Since 1994, some further categories have been distinguished. The new IUCN endangerment status categories are: extinct, extinct in wild, critically endangered, endangered, vulnerable and lower risk species (Figure 16). Additionally, categories data deficient and not evaluated can be applied as appropriate. A species is listed as threatened if it falls in the critically endangered, endangered or vulnerable categories. This is also the instruction provided in the TBFRA-2000 questionnaire if the post-1994 classification is being used. The corresponding pre-1994 categories, which should be included in the TBFRA-2000 estimates if this classification is used, are endangered, vulnerable, rare and intermediate. There may be some discrepancy in the inclusion of the old categories rare and intermediate in the TBFRA-2000 endangered categories. In some countries these are likely to contain species, which according to the new classification fall into lower risk categories (which should not be included). For instance, Lithuania classifies the national category rare to be a lower risk species category.

Figure 16. The IUCN threatened species classification (post-1994)[18]

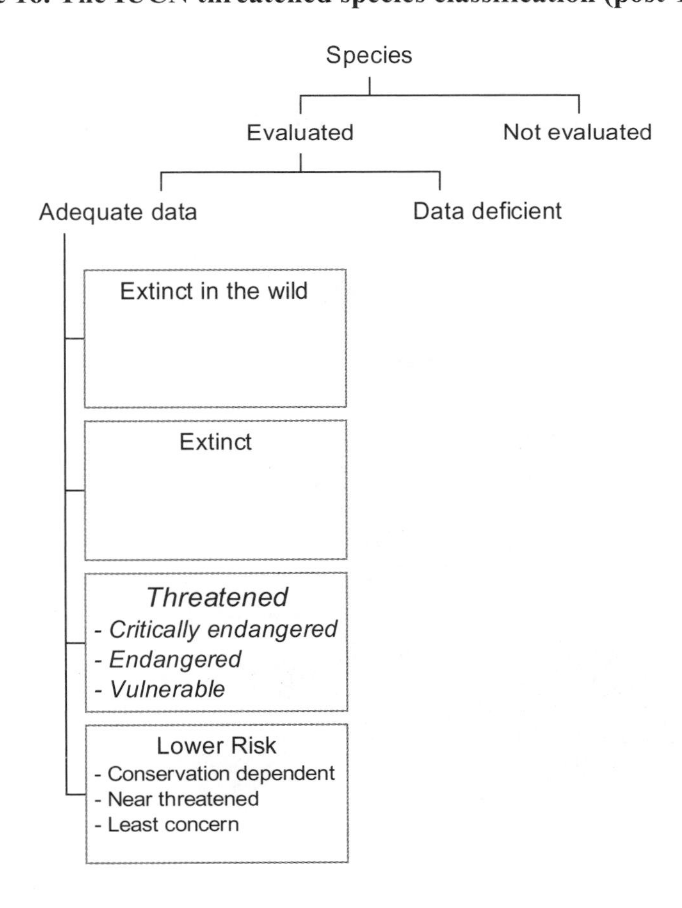

[18] The TBFRA-2000 requested data on threatened species in categories critically endangered, endangered and vulnerable.

Figure 17. Number of endangered vascular plants in different countries. Data: TBFRA-2000.

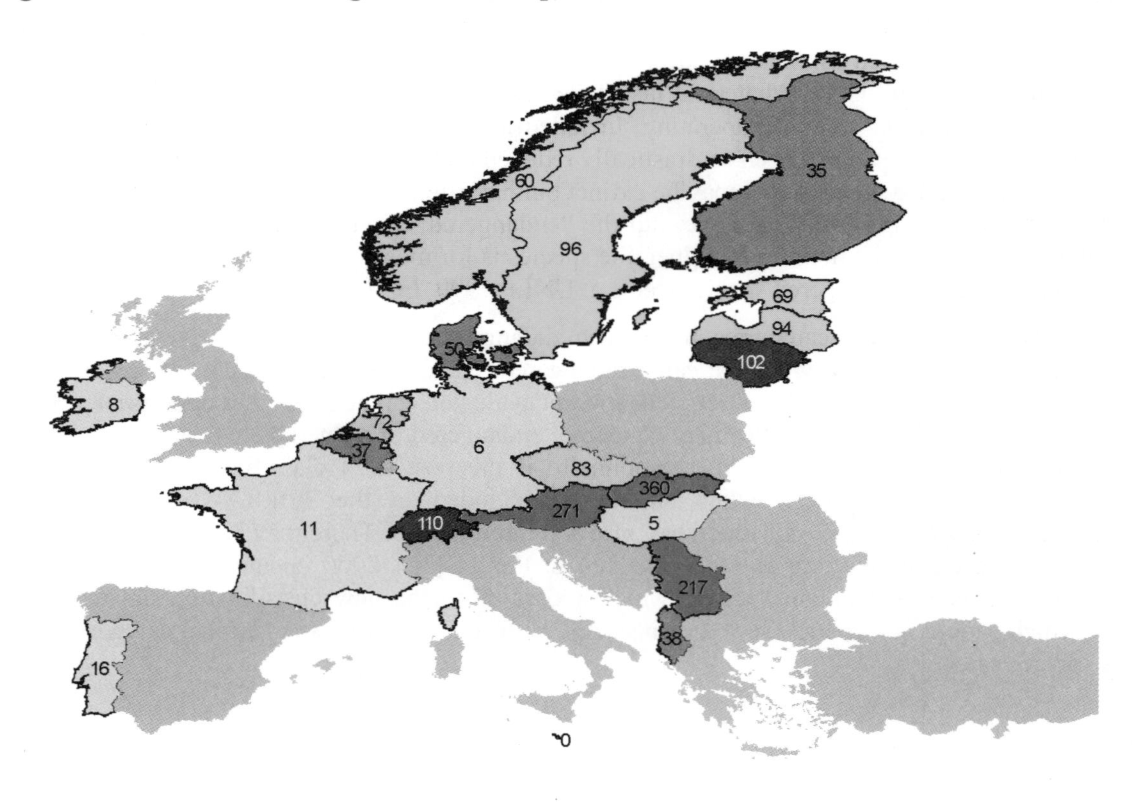

Figure 18. Number of endangered mammals in different countries. Data: TBFRA-2000.

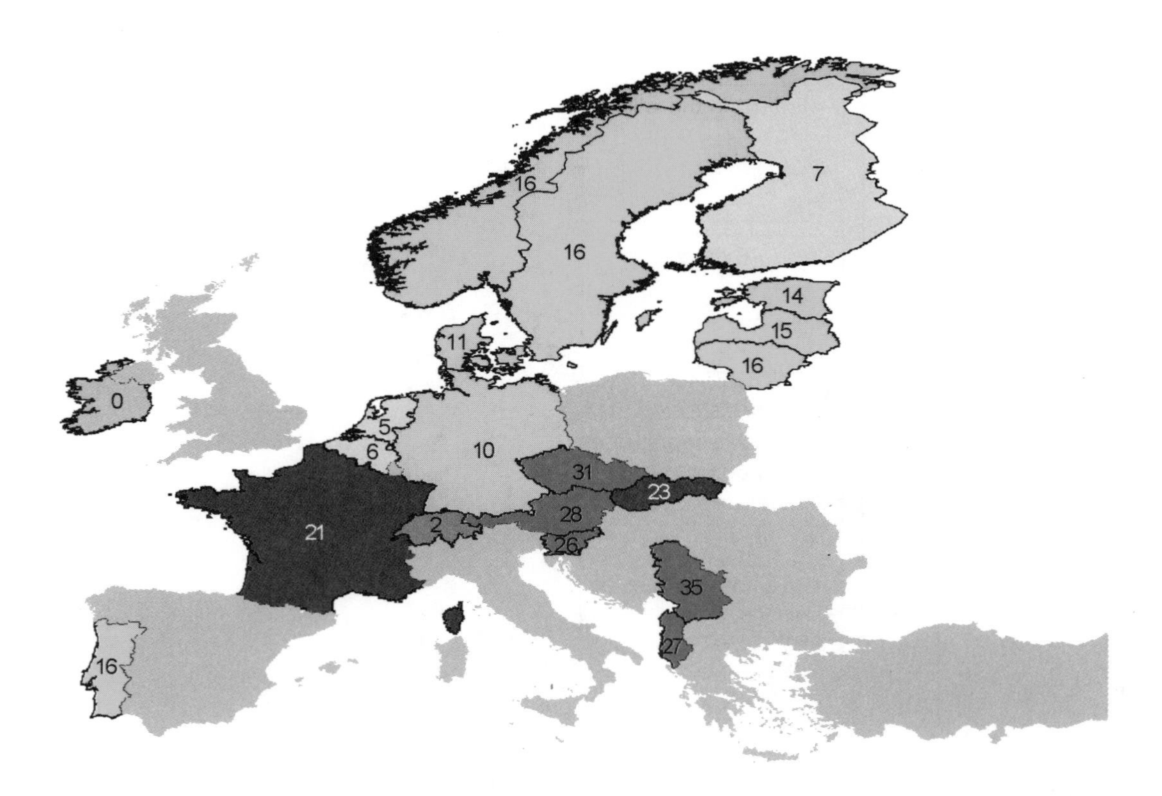

Table 7. Number of threatened species in different taxonomic groups (2000 IUCN Red List of Threatened Species)

	Mammals	*Birds*	*Other vertebrates*	*Plants*
Albania	3	3	11	0
Austria	9	3	7	3
Belgium	11	2	0	0
Bosnia and Herzegovina	10	3	8	1
Bulgaria	15	10	12	0
Croatia	9	4	23	0
Cyprus	3	3	3	1
Czech Republic	8	2	7	3
Denmark	5	1	0	3
Estonia	5	3	0	0
Finland	6	3	0	1
France	18	5	9	2
Germany	12	5	6	12
Greece	14	7	26	2
Hungary	9	8	9	1
Iceland	6	0	0	0
Ireland	5	1	0	1
Israel	14	12	4	0
Italy	14	5	16	3
Latvia	5	3	1	0
Liechtenstein	3	1	0	0
Lithuania	5	4	1	0
Luxembourg	6	1	0	0
Malta	3	1	0	0
Netherlands	11	4	0	0
Norway	10	2	8	2
Poland	15	4	1	4
Portugal	17	7	10	15
Romania	17	8	12	1
Slovakia	9	4	9	1
Slovenia	9	1	9	0
Spain	24	7	20	14
Sweden	8	2	0	3
Switzerland	6	2	4	2
The FYR of Macedonia	11	3	6	0
Turkey	17	11	37	3
United Kingdom	12	2	1	13
Yugoslavia	11	5	11	1

Note: Country totals in categories critically endangered, endangered and vulnerable (see Fig. 17) are presented for the countries of this study.

Number of endangered species in different sources

National Red Data Books exist for practically all the European countries (e.g. WCMC 1994). These are based (besides the international classifications) on the national categories for endangered species. It seems that the national Red Data Books have mainly been used as the basis for the TBFRA-information. Some discrepancies, however, become clear when the provided data are studied in detail:

- The inclusion of species listed in special national categories may increase the number of threatened species. For instance, the figures for Sweden seem to include all the red-listed species in the country, also the species in the national category "near threatened". The inclusion of this category in many cases doubles the number of species classified as threatened. Several other countries also seem to have included national categories in their estimates.
- Some countries seem to have omitted some requested categories. For instance, only the endangered category of the national Red Data Book may have been reported – but not the extinct, vulnerable or rare species.

The different classification schemes make direct comparisons difficult. Moreover, different assessments of risk, different monitoring accuracies and the general lack of knowledge for different species remain, in and between the countries under consideration.

The numbers of endangered species reported from different sources differ considerably. Some European estimates have been reviewed in the beginning of the Chapter 5. When the numbers of endangered species reported in the TBFRA-2000 are compared with the numbers of threatened species listed in the WCMC (1994) or in the IUCN 2000 threatened species data, the TBFRA-2000 estimates are far higher, independent of the species group in question (i.e. mammals, birds, other vertebrates and plant species). This is partly due to reporting of all nationally red listed species and not only the ones belonging to the requested threatened species categories. Moreover, there is a difference between a species, which is threatened at the world or European scale and which is threatened at the scale of a single country. A species may be threatened in a particular country for instance because it is at the extreme range of its natural habitat – and abundant in neighbouring countries. There is some evidence that the density of species is highest in the centre of its geographical range and declines to zero towards range margins (e.g. Hanski 1999). It also seems that the marginal populations (at least of some species) have a high temporal variability. The numbers of threatened mammals, birds, other vertebrates and plant species based on the 2000 IUCN Red list of threatened species for the countries of this study are listed in Table 7.

Data on endangered tree species in Europe further indicate the variability of information concerning endangered species. In the following, some examples are provided on the number of forest-occurring endangered tree species reported in the TBFRA-2000 and on the numbers of endangered and lower risk tree species available from the WCMC Tree Conservation Information Service. Albania has reported 21 endangered species (TBFRA-2000) compared to 1 as reported by the WCMC, Denmark 7 to 1, Slovenia 5 to 0 and Sweden 6 to 1. France, Germany, Ireland and Malta have not reported any endangered forest-occurring tree species in the TBFRA-2000, whereas WCMC lists for them 2, 9, 10 and 1 endangered and lower risk species, respectively. Poland has reported one endangered forest-occurring tree species in the TBFRA-2000 and Portugal 5 compared to 3 and 25 species respectively in the WCMC. Additionally, WCMC Tree Conservation and Information Service lists data on endangered and lower risk tree species for Bosnia and Herzegovina (1), Bulgaria (1), Croatia (2), Greece (7), Italy (2), Romania (2), Spain (37), the United Kingdom (11) and for Yugoslavia (1).

5.1.2 Total number of floral species

The total numbers of vascular plant and fern species reported in the TBFRA-2000 are contrasted with other data sources in Table 8. The following text discusses the extent and possible reasons for the deviations in different data sources.

On average there are 2,500 different vascular plant species in a country according to the TBFRA-2000[19] and 2,737 according to the other data sources. All in all considerable variation in the number of vascular plant and fern species reported in different data sources can be found. Main reasons seem to be differences in the geographical area, species groupings and all kinds of misinterpretations. For instance, it seems important to determine whether and to which extent the species number should include the species from archipelagos. This is crucial also for the determination of the extent of endemism. In the following these comparisons are summarised in a greater detail.

There is no real systematic difference between the reported figures, but the estimates for each single country vary in both directions – and in some cases considerably. For some countries the figures reported in the TBFRA-2000 are smaller and for some countries larger than the data obtained from other sources. At best, the differences are "small" – which mean less than 50 species differences for eight of the 29 reported countries. The 50 species represent about 2% of the European average per country. Generally this means that the total numbers of (vascular) plant species provide an indication of the variation of species richness in different European countries. Their credibility as such for monitoring the status of biodiversity is, however, questionable. In addition, the knowledge over the total numbers of mosses and lichens is moderate compared with that for vascular plants or fern species.

Ferns

The average number of fern species in a country is 56 according to the TBFRA-2000 and 62 according to the WCMC (1994). The number of countries for which the data are available, however, varies. Almost identical numbers in the two data sources have been reported for seven countries (Austria, Finland, France, Hungary, Latvia, Poland and Switzerland). In eight countries the number of fern species reported in the TBFRA-2000 is smaller and in five countries larger than the one from WCMC (1994).

The largest differences in the reported species numbers are in Portugal (+49 species in the TBFRA-2000), Estonia (+38 species), Ireland (+22 species), Yugoslavia (-21 species) and Belgium (-21 species). In Portugal the difference may lay in the definition of the geographical area, i.e. in inclusion or exclusion of archipelagos. Further recent documentation supports the figures reported in the TBFRA-2000 for Ireland, the UK, Turkey, Slovenia, Latvia, Israel and Hungary (see the references from the Tab. 8). Israel has reported only the number of wild fern species (25). A further 75 species are cultivated in the country.

---

[19] This average does not include the remarkably small number of 12 vascular plant species reported for Malta. Also, the number of countries for which the data of these figures are available is not the same.

Table 8. Total numbers of vascular plants and fern species reported from different sources[20]

	Total number of vascular plant species			Total number of fern species	
	TBFRA-2000	Comparison 1	Comparison 2 [c]	TBFRA-2000	WCMC (1994)
Albania	3,250	3,965 [a]	3,200	-	45
Austria	2,931	2,873 [a]	2873	54	66
Belgium	1,270	1,415 [a]	-	29	50
Bosnia and Herzegovina	-	3,572 [b]	-	-	-
Bulgaria	-	3,583 [a]	3,550 – 3,750	-	52
Croatia	3,871	4,283 [c]	-	75	-
Cyprus	1,910	1,682 [b]	-	20	-
Czech Republic	2,969	2,500 [a]	2,520	68	-
Denmark	1,298	1,200 [a]	-	48	50
Estonia	1,437	1,448 [c]	1,560	42	4
Finland	1,244	1,102 [b]	-	59	58
France	4,564	4,630 [b]	4,700	110	110
Germany	3,236	3,203 [c]	>3,240	83	73
Greece	-	4,992 [b]	5,500	-	71
Hungary	2,346	2,214 [b]	2,346	60	58
Iceland	495	483 [c]	-	17	36
Ireland	1,230	950 [b]	1,309	78	56
Israel	2,781	2,780	-	25	-
Italy	-	5,820 [c]	5,463	-	106
Latvia	1,669	1,658 [a]	1,678	48	48
Liechtenstein	1,639	1,410 [b]	-	35	-
Lithuania	1,354	1,609 [a]	1,796	21	-
Luxembourg	-	1,246 [b]	-	-	43
Malta	12	914 [b]	-	0	11
Netherlands	1,404	1,221 [b]	-	32	48
Norway	1,343	1,310 [a]	-	-	61
Poland	2,335	2,300 [a]	2,300	69	62
Portugal	4,663	3,150 [a]	3,200	114	65
Romania	-	3,350 [a]	3,700	-	62
Slovakia	2,491	3,124 [c]	4,178	63	-
Slovenia	3,100	3,216 [c]	3,100	75	75 [c]
Spain	-	5,048 [b]	7,500	-	114
Sweden	1,932	1,900 [a]	-	50	60
Switzerland	2,644	2,696 [b]	2,696	84	87
The FYR of Macedonia	-	3,500 [c]	-	-	-
Turkey	8,950	8,579 [b]	8,950	78	85
United Kingdom	1,640	1,623 [b]	1,400	80	70
Yugoslavia	-	4,282 [c]	-	57	78

[20] Total number of vascular plants in the TBFRA-2000 has been calculated as a sum of the total number of tree species and total number of other vascular plants. For the details of the data see the table reference note in the next page.

a) data from the WCMC (1994).

b) data from European Commission (1995).

c) data from other, mainly national, sources:

- *Albania: Comparison 2: Anon. (1998a). Includes continental vascular plant species.*
- *Austria: Comparison 2: Anon. (1997a) Includes ferns and angiosperms.*
- *Bulgaria: Comparison 2: Bojinov et al. (1998).*
- *Croatia: Comparison 1: Anon. (1999a) National biodiversity and landscape strategy for Croatia.*
- *Czech Republic. Comparison 2: Plesnik and Roudna (1999) estimate the number of higher plant species to 2,520.*
- *Estonia. Comparison 1: Peterson (1994). Comparison 2: Peterson et al. (1998).*
- *France. Comparison 2: Anon. (1997b).*
- *Germany. Comparison 1: Anon. (1995). Comparison 2: Anon. (1998c).*
- *Greece. Comparison 2: Legakis and Spyropoulou (1998). The figure includes flowering plants.*
- *Hungary: Comparison 2: (Anon. 1998d). A total of 60 fern species and 2,346 vascular plant species (2,343 Angiospermae and 3 Gymnospermae).*
- *Iceland: Comparison 1: Anon. (1992).*
- *Ireland. Comparison 2: Anon. (1999b). Beside the 1,309 vascular plant species 78 fern species have been reported.*
- *Italy: Comparison 1: OECD (1994). Comparison 2: ENEA (1998).*
- *Israel. Comparison 1: Gabbay (1997). The total of 2,780 wild plant species includes also exotics and 500 of the species are wetland, moist land and marine habitat plants. There are also 25 wild and 70 cultivated fern species in Israel.*
- *Latvia. Comparison 2: Kabucis et al. (1998). 1,678 vascular plant species and 49 fern species in Latvia.*
- *Lithuania. Comparison 2: Anon. (1998e) The total number(1,796) for aquatic and terrestrial plant species*
- *Poland. Comparison 2: Anon. (1997c).*
- *Portugal. Comparison 2: Anon. (1998f). There are 3,200 continental vascular plant species in Portugal. Additionally there are 1,200 vascular plant species in the archipelago of Madeira and Selvagens, and 750 vascular plant species in Azores.*
- *Romania. Comparison 2: Anon. (1998g). About 3,700 higher plant species.*
- *Slovakia: Comparison 1: Comparison 1: Klinda (1998). Comparison 2: Marhold and Hindak (1998).*
- *Slovenia: Comparison 1: Trpin and Vres (1995). Comparison 2: Anon. (1997d). The taxonomic groups Pteridophyta (ferns) and Spermatophyta (vascular plants) are estimated to have 75 and 3,100 species, respectively.*
- *Spain. Comparison 2: Anon. (1997e). The total of 7,500 vascular plants includes the Iberian and Balearian species, but not the species of Canary Islands. In the Canary Islands there are 1.992 vascular plant species.*
- *Switzerland. Comparison 2: SAEFL (1998). The total of 2,696 species includes higher plants and ferns.*
- *The FYR of Macedonia: Comparison 1: (Matevski et al. 2000). National report on biodiversity records a total of 3,500 higher plant species. The figure does not include moss, algae and fungi.*
- *Turkey. Comparison 2: Anon. (1998h). Reported numbers of vascular plants (8,950), ferns (78), seed plants (8,869), Gymnosperm (22) and Angiosperm (8,850).*
- *The United Kingdom. Comparison2 : Anon. (1998 i). There are 1,400 flowering plant and 80 fern species.*
- *Yugoslavia: Comparison 1: Stevanovic and Vasic (1995).*

Vascular plants

Generally, for the number of vascular plants the comparisons can be summarised as follows (see Table 8):

- Countries with no TBFRA-2000 data: The estimates are complemented for nine countries for which the data are not available from the TBFRA-2000, namely for Bosnia and Herzegovina, Bulgaria, Greece, Luxembourg, Romania, Spain, the FYR of Macedonia and Yugoslavia. Obviously the figures may not be directly comparable, but still provide an indication of the number of vascular plant species at the country level.

- Countries with only slight differences in different data sources: For eight countries the difference between the reported numbers in at least two different data sets is less than 50 species (Estonia, Germany, Iceland, Latvia, Norway, Poland, Sweden and the United Kingdom).

- Countries with some variation in different data sources: The numbers of species reported in the different data sources vary. The reasons for the variations are sometimes difficult to trace, but sometimes also possible to identify. The following aspects have been identified:

  - More than a 10% but not a "very large" difference between the vascular plant species in the TBFRA-2000 and in another data source are reported for Belgium, Croatia, Cyprus, Czech Republic, Finland, Hungary, Ireland, Liechtenstein, Lithuania and the Netherlands. In Belgium there may be some difficulties to combine and interpret the results for the Flemish and Walloon regions. In Croatia both reported figures originate from recent documentation, i.e. 3,871 species in the TBFRA-2000 and 4,283 in the new national biodiversity report (Anon. 1999a). For The Czech Republic, Plesnik and Roudna (1999) report 2,520 higher plant species, WCMC (1994) 2,500 and TBFRA-2000 2,969 vascular plant species. In Lithuania, the number reported in the TBFRA-2000 is smaller (1,354) than the one (1,796) from the recent national biodiversity report (Anon. 1998e). The latter includes " aquatic and terrestrial plant species", which may include not only vascular plants. In Hungary and Ireland the most recent data sources provide fairly similar estimates.

  - Differences in the geographical area from which the number of species is reported are likely in Portugal, and also Albania and Spain. In Portugal, the TBFRA-2000 species number (4,663) includes the species from the Archipelago of Madeira and Selvageno as well as the Azores, whereas the smaller estimates from other sources (3,150 – 3,200) contain only the continental vascular species (Anon. 2000). In Albania the new national biodiversity report states that the 3,200 vascular plant species are solely continental species. The WCMC (1994) figure of 3,965 species is likely to include further species from the archipelagos. In Spain the estimates range between 5,048 – 7,500 species, depending on the inclusion of only continental species, of part of the archipelagos or of the species from the archipelagos inclusive of the Canary Islands.

  - Differences in species groupings seem clear in some cases. For instance the national figures reported for Austria (Anon. 1997a) and Switzerland (SAEFL 1998) include both vascular plants and ferns. In some data sources terms such as higher plant species, higher plants and ferns, flowering plants, terrestrial and marine wild plant species inclusive of exotics, and aquatic and terrestrial plant species have been used. The reported figures may be similar, larger or smaller than those obtained from other sources for vascular plant species of the country.

  - Differences may also be due to misunderstanding and misinterpretation of the questions posed, of the national databases and figures and of the data received as replies to questionnaires. For instance, in Slovakia the number of vascular plant species reported from different sources are 2,491 (TBFRA-2000), 3,124 (Klinda 1998) and 4,178 (Marhold and Hindak 1998). No clear reason for the discrepancies was found.

5.1.3 Total number of faunal species

The total numbers of mammals, birds and other vertebrate species reported in the TBFRA-2000 are contrasted with other data sources in Table 9. The following text discusses the extent and possible reasons for the deviations in different data sources.

Mammals

On average there are 76 (TBFRA-2000) or 67 (WCMC 1994) mammal species in a European country. In 21 countries the TBFRA-2000 figure is larger and in seven countries smaller than the number of mammals reported in the WCMC (1994). Only in Estonia is the reported number the same. The differences in the reported numbers of these two data sources are within ± 5 species in eight countries, within ± 10 species in a further six countries and over 10 species in 14 countries. The figure of a single country seems to vary providing if only regularly residing species or the total recorded number are given. Further reasons for the varying figures may be the ones listed already for the case of flora: Differences in the geographical area, species groupings and all kinds of misinterpretations. Generally the group of mammals is well known.

Birds

The total number of bird species reported by the countries seems to refer to the total number inclusive of migratory species, to the number of breeding species or to a number between these two. These differences become clear when the TBFRA-2000 data are compared with other data sources, which differentiate between the total and breeding bird species (see Table 9). For most countries the figures are clearly close to one of the two figures. Since Europe is a seasonal home and crossroads for huge populations of migratory birds, the differences between the total and breeding bird species number are not negligible. Based on the WCMC (1994) data a European country can record, on average, a total of 406 bird species of which 203 are breeding. The average number of bird species calculated from the TBFRA-2000 data is 305 per country. Since the countries have not necessarily reported the same element, the TBFRA-2000 figure for bird species should be interpreted very carefully. The countries have reported as follows:

- Breeding bird species have been reported by 15 countries (Austria, Belgium, Croatia, Denmark, Finland, France, Germany, Iceland, Italy, Liechtenstein, Luxembourg, the Netherlands, Norway, Sweden and Switzerland), 10 of which belong to the European Union.
- The total number of observed bird species including the migratory species has been reported by 14 countries, mostly in Eastern Europe (Albania, Bulgaria, Cyprus, Czech Republic, Hungary, Ireland, Israel, Latvia, Lithuania, Malta, Slovakia, Slovenia and Turkey).
- A figure between the total and breeding bird species number has been reported by Portugal, Spain and the UK. The figure for the UK (390 bird species) is likely to contain 210 breeding and 188 wintering terrestrial bird species (Anon. 1998h). Additionally 188 marine bird species have been recorded in the country. The reasons for the differences in the Estonian estimates were not found (549 bird species against other estimates of 213 - 332 species).

Reptiles, amphibians and fishes

The number of other vertebrates (as recorded in the TBFRA-2000) is also commonly divided between reptiles, amphibians and (freshwater) fish species in international reporting. Information on the number of reptiles and amphibians is easily available, whereas data on fish species are more often lacking (e.g. WCMC 1994). Thus, the comparisons between the total number of other vertebrate species is meaningful only for a few countries. In ten countries the TBFRA-2000 data and another data source have a difference of maximum ± 10 species. These countries are Bulgaria, the Czech Republic, Hungary, Israel, Latvia, Lithuania, Poland, Slovakia, Slovenia and the United Kingdom. Generally, the reported numbers from different data sources differ to some extent without a clear and obvious reason. In some TBFRA-2000 figures fish species are likely to be missing and in others marine fish species may have been included beside the freshwater ones.

Table 9. Total numbers of mammals, birds and other vertebrate species reported from different sources[21]

	Mammals			Birds				Other vertebrates					
	Total number			Reported	Total	Breeding	Reported	Total	Total	Total [22]	Reptiles	Amphibians	Freshwater fish
Data	TBFRA-2000	WCMC 1994	Other[a]	TBFRA-2000	WCMC 1994	WCMC 1994	Other[a]	TBFRA-2000	Other [a]	WCMC 1994	WCMC 1994	WCMC 1994	WCMC 1994
Albania	84	68	70(84)	320	306	230	323	365	115+249	-	31	13	-
Austria	96	83	82	250	414	213	239	33	101	-	14	20	-
Belgium	57	58	68	167	429	180	-	65	-	-	8	17	-
Bulgaria	94	81	94	383	374	240	383	259	259	-	33	17	-
Croatia	100	-	-	232	-	224	-	600	-	-	-	-	-
Cyprus	25	21	-	365	347	79	-	26	-	-	23	4	-
Czech Republic	76	-	86 or 87	396	-	199	186/390 or 220	88	97	-	-	-	-
Denmark	48	43	-	185	439	196	-	58	-	-	5	14	-
Estonia	65	65	64	549	330	213	332	87	92	46	5	11	30
Finland	66	60	-	240	425	248	-	104	-	76	5	5	66
France	119	93	100	284	506	269	-	73	-	-	32	32	-
Germany	86	76	98	255	503	239	273	35	101	-	12	30	-
Greece	-	95	116	-	398	251	422	-	189	164	51	15	98
Hungary	83	72	83	370	363	205	-	111	112	39	15	17	7
Iceland	6	11	-	100	316	88	-	6	-	-	0	0	-
Ireland	29	25	-	420	417	142	-	19	-	39	1	3	35
Israel	116	92	116	511	500	180	511/204	152	152	-	-	-	26
Italy	105	90	97 or 118	230	490	234	406 or 473	140	58 reptiles, 38 amphibians	-	40	34	-

[21] For the details of the data see the table reference note at the end of the table.

[22] The total number of other vertebrates in the WCMC (1994) has been calculated as a sum of reptile, amphibian and freshwater fish species.

	Mammals			Birds				Other vertebrates					
	Total number			Reported	Total	Breeding	Reported	Total	Total	Total [22]	Reptiles	Amphibians	Freshwater fish
Data	TBFRA-2000	WCMC 1994	Other[a]	TBFRA-2000	WCMC 1994	WCMC 1994	Other[a]	TBFRA-2000	Other [a]	WCMC 1994	WCMC 1994	WCMC 1994	WCMC 1994
Latvia	69	83	69	320	325	217	320	118	115+3	129	7	13	109
Liechtenstein	56	64	-	145	235	124	-	39	-	-	7	10	-
Lithuania	70	68	70	321	305	202	323	120	119	-	7	13	-
Luxembourg	-	55	-	136	289	126	-	-	-	-	7	14	-
Malta	-	22	-	360	395	26	-	-	-	-	8	1	-
Netherlands	65	55	-	172	456	191	-	23	-	-	7	16	-
Norway	76	54	57	220	453	243	-	35	-	-	5	5	-
Poland	93	79	90	360	421	227	365-370/229	139	143	-	9	18	-
Portugal	70	63	68	350	441	207		46	35 reptiles, 18 amphibians	74	29	17	28
Romania	-	84	102	-	368	247	364	-	241	131	25	19	87
Slovakia	85	-	86	335	-	209	352	111	108	-	-	-	-
Slovenia	88	69	75	361	361	207	360	144	137	-	21	-	98
Spain	118	82	-	368	506	278	-	149	-	128	53	25	50
Sweden	69	60	-	245	463	249	-	161	-	-	6	13	-
Switzerland	57	75	-	205	400	193	-	89	-	-	14	18	-
The FYR of Macedonia	-	-	78	-	-	-	330	-	99	-	-	-	-
Turkey	132	116	132	450	418	302	450	635	635	>272	102	18	>152
The UK	48	50	48	390	590	230	210 (breeding terrestrial)	50	50	50	8	7	36
Yugoslavia	96	98	-	382	-	-	-	180	-	-	-	-	-

a) data from other, mainly national, sources:

- ***Albania (Anon. 1998a). A total of 70 mammal species are regularly residing in the country, 84 is the total number recorded. The number for other vertebrates includes 36 reptile, 15 amphibian and 64 freshwater fish species. Additionally there are 249 marine fish species found in the country.***
- ***Austria (Anon. 1997a). Other vertebrates (101 species) include 16 reptile, 21 amphibian and 64 fish species.***

- *Belgium: Anon (1998j). 68 mammal species have been reported both for the Flemish and Walloon regions. 19 reptile &hibian and 161 summer bird species have been reported for the Flemish region. The Walloon region has reported 339 bird and 53 fish species.*
- *Bulgaria: (Anon. 1998b). Other vertebrates include 36 reptiles, 16 amphibians and 207 fresh water and Black Sea fish species (total to 259 species).*
- *Czech Republic: (Plesnik and Roudna 1999). Total number of mammals equals to 87 according to EC (1995) and to 86 (Plesnik and Roudna 1999). In EC (1995) 220 bird species have been reported and according to Plesnik and Roudna (1999) there are 186 recently regularly nesting bird species in the country and a total of 390 species have been detected since 1800. The 97 other vertebrate species include 11 reptile, 21 amphibian and 65 fish & lamprey species.*
- *Estonia: (Kull 1999). The 92 other vertebrate species include 5 reptile, 11 amphibian and 76 fish species.*
- *France: Anon. (1997b).*
- *Germany (Anon. 1998c). The number of bird species (273) includes only incubating birds. Other vertebrates include 12 reptile, 19 amphibian and 70 freshwater fish and cyclostome species. The number of marine fish and cyclostome species totals to 115.*
- *Greece: Legakis and Spyropoulou (1998). Other vertebrates include 59 reptiles, 20 amphibians and 110 freshwater fish species. There are also 447 marine fish species in Greece.*
- *Hungary: (Anon. 1998d). Other vertebrates include 15 reptiles, 16 amphibians and 81 fish species.*
- *Israel: Gabbay (1997). The mammal species include 106 terrestrial mammal species and an estimate of 10 marine mammal species. The total number of bird species (511) includes 204 breeding species. Other vertebrates include 97 reptile, 7 amphibian, and 48 freshwater fish species (of which 12 are cultivated). Additionally there are 410 Mediterranean and 1,270 Red Sea fish species.*
- *Italy: ENEA (1998). The total number of Italian mammal species is 118, 93 of which are protected. The number of bird species equals to 473 and 58 reptiles and 38 amphibians have been reported (ENEA 1998).OECD (1994) records 97 mammal and 406 bird species.*
- *Latvia: (Kabucis et al. 1998). Other vertebrates include 7 reptiles, 13 amphibians and 95 fish species. There are also 3 lamprey species in the country.*
- *Lithuania: (Anon. 1998e). Other vertebrates include 7 reptiles, 13 amphibians and 99 fish species.*
- *Norway: OECD (1993).*
- *Poland (Anon. 1997c). 365-370 bird species occur in Poland, of which 229 breed. Other vertebrates include 9 reptiles, 18 amphibians and 116 fish species (of which 23 alien).*
- *Portugal: Anon. (1998f).*
- *Romania: (Anon. 1998g). 364 bird species include both nesting and migratory species. Other vertebrates include 30 reptile, 20 amphibian and 191 fish species.*
- *Slovakia: Anon. (1998k). The number of other vertebrates has been calculated as the sum of reptiles (12 species), amphibians (18 species) and fishes (78 species).*
- *Slovenia: (Anon. 1997d). Other vertebrates include 22 reptile, 20 amphibian and 95 fish species. These are the terrestrial species in Slovenia (i.e. not sea).*
- *Spain: (Anon. 1997e). Total number of vertebrate species equals to 635 in Spain (Iberia and Baleares) and 122 in the Canary Islands.*
- *The FYR of Macedonia: Matevski et al. (2000). The 99 other vertebrates include 31 reptiles, 13 amphibians and 55 fish species.*
- *Turkey: (Anon. 1998h). Other invertebrates include 21 amphibians, 106 reptile and 508 fish species (inclusive marine fish species).*
- *The UK: Anon. (1998i). There are 210 breeding and 180 wintering terrestrial bird species and 188 marine bird species in the UK. Other terrestrial vertebrates include 6 reptile, 6 amphibian and 38 freshwater fish species. There are a further 300 marine fish species.*

5.2 Forests as a Species Habitat

The role of forests and other wooded land as a habitat for different species is estimated to be crucial at a global scale. "The forests are arguably the single most important repository of global biodiversity ..." (Kapos and Iremonger 1998). The knowledge of the world species diversity as such is incomplete[23] and there is hardly any credible estimate of the number of species residing in forests. The edges between forests and other land use forms constitute a further area for special attention, because the number of species found in such transformation zones is often exceptionally high.

European forests have been subject to anthropogenic changes for a long period of time, the forest cover is only a fraction of the original one and qualitative changes have taken place. Are forests and other wooded land still an important habitat for different flora and fauna species in Europe? How important? Are forest-related species generally more endangered than species related to other habitats? Which species groups are particularly endangered? These kind of data assist in setting conservation priorities and in selecting which kinds of habitats and which organisms are proportionally the most endangered. The TBFRA-2000 data allow a rough estimation[24] on:

- How many species are related to forest ecosystems in different European countries.
- If the forestry-related species are proportionally more endangered than the species in other habitats.
- Which taxonomic groups are particularly forest-dependent or endangered in forest ecosystems.

5.2.1 Forest-occurring floral species

In total, especially Central and Eastern European countries report a fairly high percentage of different plant species occurring in forests (e.g. Austria, Yugoslavia, Czech Republic, Slovakia, Poland, Lithuania and Latvia), be it vascular plats, ferns, mosses or lichens. This may be due to a large proportion of forests of the land area and to structurally diverse forests. If area alone were to count, then the contribution of forests for plant species richness would be also high in the Nordic countries.

Tree species

Most of the tree species found in Europe are forest-occurring. Species found solely in the parks, hedges or possibly in arboreta but not in forests have been reported by 15 countries. The countries (and the numbers of trees not found in forests) are: Albania (210), Belgium (20), Bulgaria (3,540), Cyprus (35), Denmark (35), Estonia (12), France (32), Germany (3), Hungary (61), Ireland (45), Israel (10), Lithuania (52), Malta (1), the Netherlands (4) and Sweden (2). Especially the figures for Albania and Bulgaria are likely to include species in arboreta. Spain has reported a total number of 8,500 tree species found in the country, which also seems to include species in arboreta (or refers to the total number of vascular plants).

Vascular plants

The number of forest-occurring vascular plant species varies from below 200 species to over 2,000 species in different countries in Europe (Figure 19). The countries with the largest numbers are located more in the South and East than in the West and North. The countries with the smallest numbers have either a small forest area (e.g. Ireland, Portugal, Belgium and the Netherlands) or are located in northern Europe (e.g. Finland, Estonia and Sweden). The highest numbers have been recorded for southern and eastern European countries.

---

[23] The global species number may be about 13 – 14 million species of which 1.75 million have been described scientifically (Jaakkola 1998).

[24] The results are somewhat limited, and restricted to the countries where the data are available. The considerations to carefully interpret the number of reported species apply, and it should also be noticed that an inclusion of a species into forest-occurring or not may in some cases be highly subjective. Since the data on forest-occurring species have seldom been collected and reported prior to the TBFRA-2000 assessment, the material provides interesting new data but also allows little possibility for verification.

Countries where at least half of the vascular plant species are found in forests are Albania, Norway, Israel, Lithuania, Slovakia, Malta, Cyprus and Yugoslavia (Figure 20). In most other countries 10 – 30% of the vascular plant species of a country reside in forest habitats. Vascular plants tend to be more endangered in non-forest than in forest habitats. For instance, in France and Germany about 16-19% of vascular plants are found in forests, but only 2-4% of the endangered vascular plants are forest-occurring. Only Estonia and Cyprus are exceptions in this respect.

Ferns

At least half of the fern species are forest-occurring in most countries from which the data are available (Figure 21). The highest proportions of forest-occurring ferns are located in the eastern part of Europe. In only four countries the majority of fern-species reside outside forest ecosystems, namely in France, Ireland, Portugal and Switzerland. It also seems that the fern species related to forest ecosystems are less endangered than fern species in other ecosystems. In only four countries the situation is opposite, namely in Germany, Hungary, Portugal and Sweden.

Mosses and lichens

Generally, the data on moss and lichen species are relatively limited. The reported percentages of forest-occurring moss and lichen species as a proportion of the total number of these species varies between 20 – 100 % and 10 – 100%, respectively, between different countries. As in the case of ferns the percentage of forest-occurring species tends to increase towards the East in Europe. Moss species are somewhat less endangered in forest ecosystems compared to other habitats and also less endangered than forest-occurring lichen species. At least 50 forest-occurring lichen species are endangered in 8 countries out of the 13 reported ones in Europe. Furthermore, in 7 countries (out of 16) at least 50 forest-occurring moss species are endangered.

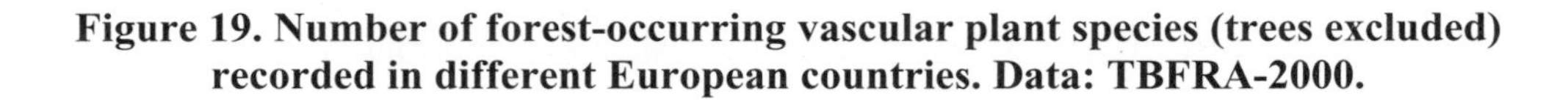

Figure 19. Number of forest-occurring vascular plant species (trees excluded) recorded in different European countries. Data: TBFRA-2000.

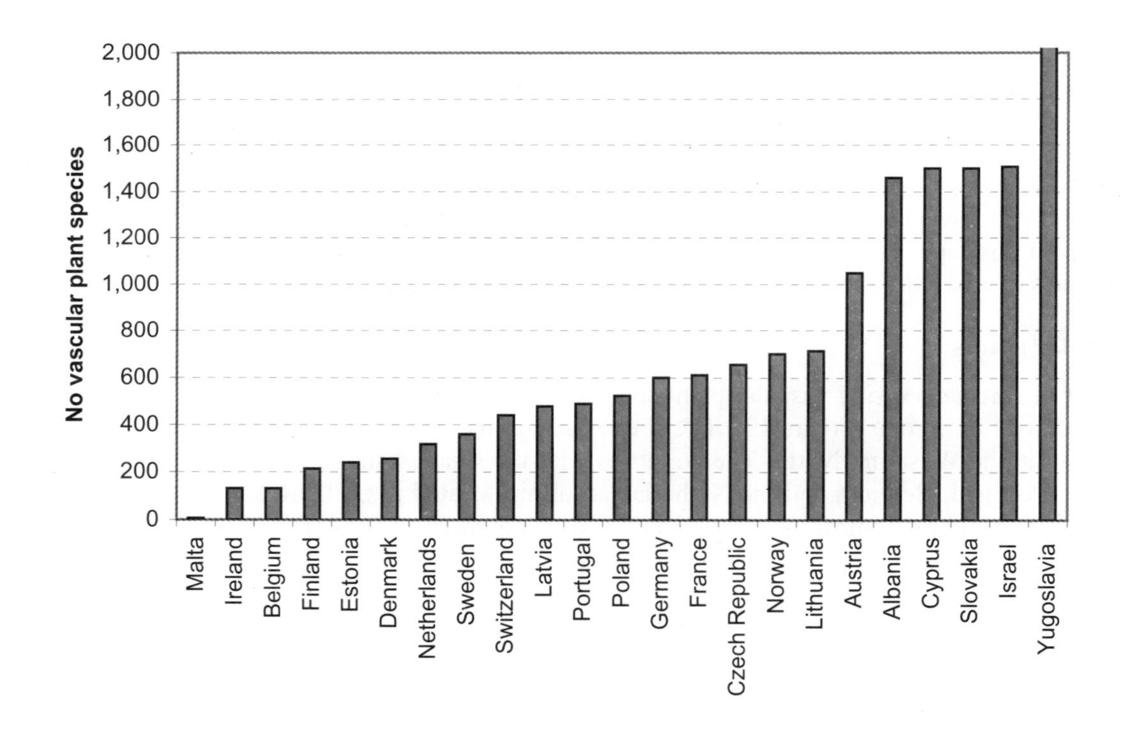

Figure 20. All forest-occurring vascular plants and the endangered ones as percentage of the total and endangered vascular plants, respectively. Data: TBFRA-2000.

Forest-occurring vascular plants % total (endangered)
100.0
80.0
60.0
40.0
20.0
0.0
Portugal
Belgium
Ireland
France
Switzerland
Finland
Estonia
Sweden
Germany
Denmark
Poland
Netherlands
Czech Republic
Latvia
Austria
Albania
Norway
Israel
Lithuania
Slovakia
Malta
Cyprus
Yugoslavia
Hungary
Iceland
Forest-occuring % total
Endangered forest-occuring % total endangered

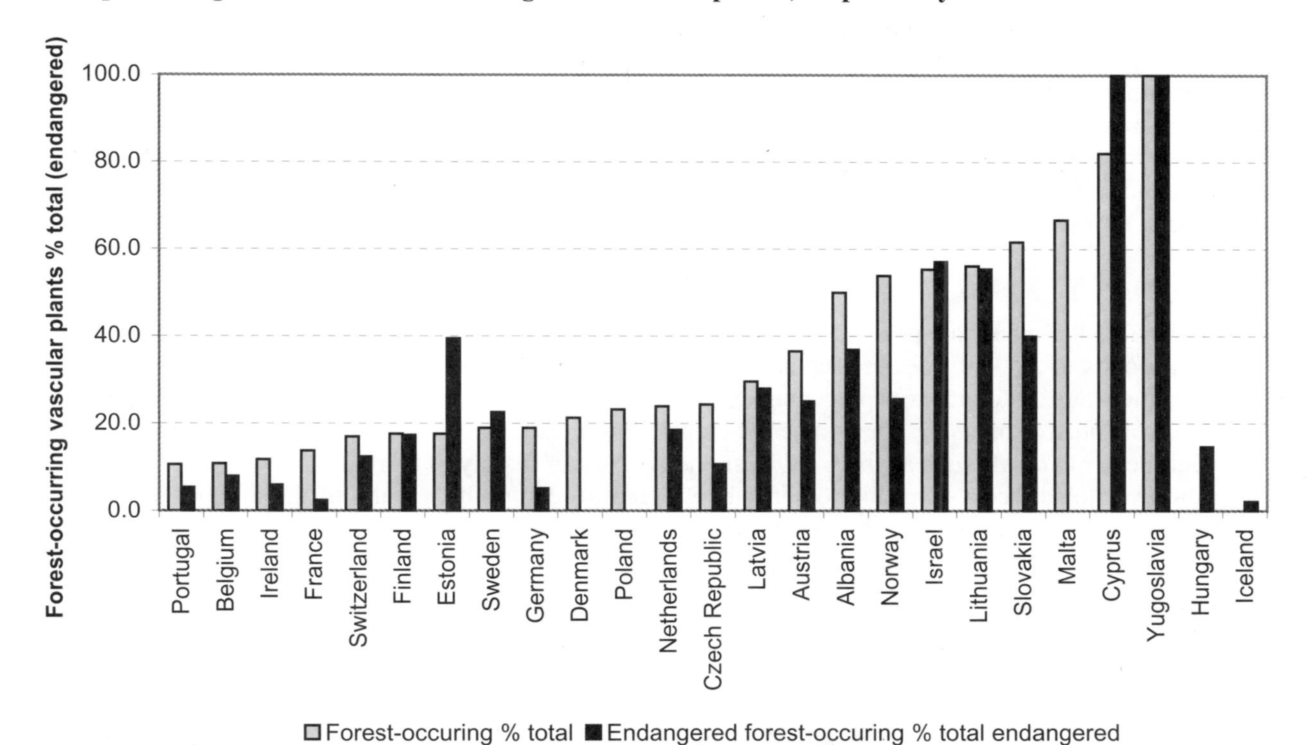

Figure 21. All forest-occurring ferns and the endangered ones as percentage of the total and endangered ferns, respectively. Data: TBFRA-2000.

Forest-occurring ferns % total
100.0
80.0
60.0
40.0
20.0
0.0
France
Ireland
Portugal
Switzerland
Finland
Denmark
Estonia
Belgium
Netherlands
Germany
Austria
Slovakia
Hungary
Sweden
Israel
Latvia
Lithuania
Czech Republic
Cyprus
Yugoslavia
Forest-occurring ferns % total
Endangered forest-occurring % of total endangered

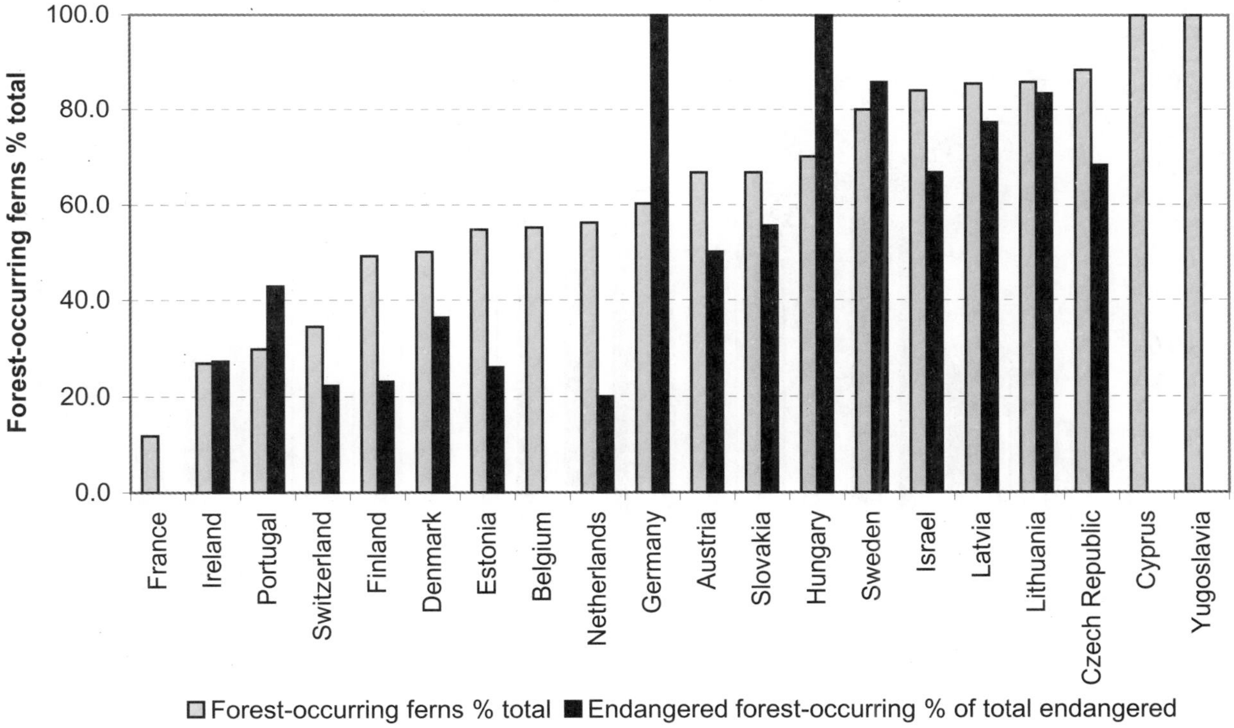

5.2.2 Forest-occurring faunal species

There seems to be a slight tendency for the larger animals, particular mammals and birds, to be proportionally more endangered than the smaller creatures in the TBFRA-countries (TBFRA-2000 Main Report). The absolute numbers of endangered species of different animal groups may in some countries be almost equally large, even though the total number of species clearly varies (e.g. number of mammal and insect species). The reason may lie in the fragmentation of forest ecosystems into patches, which are too small for larger mammals, but still suitable for smaller species - or simply be due to the fact that mammal species are better understood, than insects, for example.

Mammals

At least 50% of the mammal species reside in forest ecosystems in the most European countries (Figure 22). This corresponds to 20 – 96 species, depending on the country. A total of 15 countries report a percentage of 60% or higher, and 5 countries a percentage of at least 80% for the forest-occurring mammal species. The importance of forest habitats for the number of mammal species seems to increase towards the East in Europe. Large European mammals such as the wolf, bear and lynx occur mainly in Northern and Eastern Europe. At the European level forest ecosystems and other habitats seem to contribute equally to the number of endangered mammal species. In Western Europe 50 and in Eastern Europe 35 mammal species are endangered (UNEP 1999). In single countries the results vary.

Butterflies and moths

The endangered butterfly species are most often related to habitats, which depend upon the traditional agricultural practices (Van Swaay and Warren 1999) – and not so much on forests. This is to some extent supported also by the TBFRA-2000 data. Generally, forest-occurring butterflies and moths tend to be less endangered than the species related to other habitats – except in Albania, Sweden, The Netherlands and Estonia.

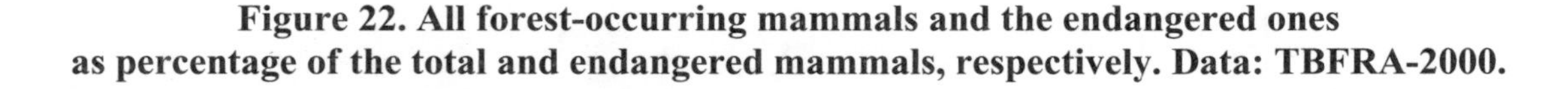

Figure 22. All forest-occurring mammals and the endangered ones as percentage of the total and endangered mammals, respectively. Data: TBFRA-2000.

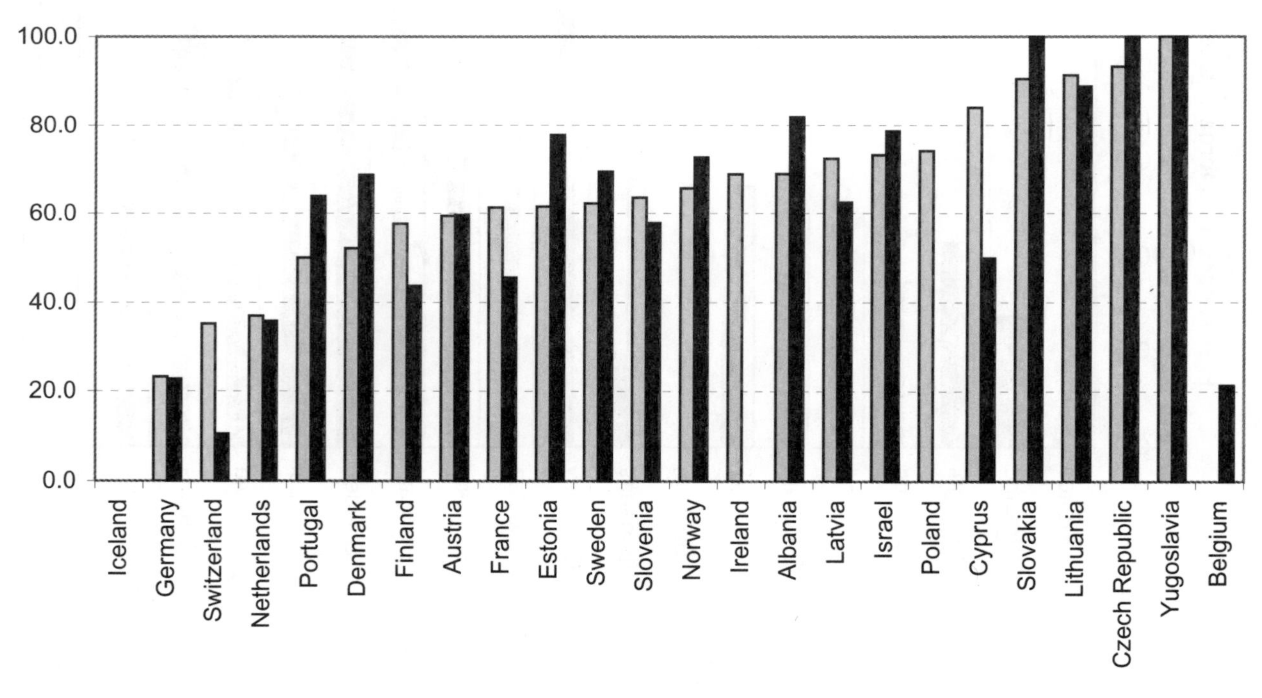

Birds

Generally, birds seem to be less dependent on forests as a habitat than for instance mammals in Europe. In most countries 20 – 60% of bird species are forest occurring. Only in Israel, the Czech Republic and Yugoslavia is the percentage higher. The number of forest-occurring bird species in Europe increases towards the North and East of Europe (Figure 23). For instance Norway, Sweden and Finland, the Baltic countries (excluding Estonia), as well as the Czech Republic, Slovakia and Austria all report more than 100 forest-occurring bird species whereas only 53 species have been reported by Germany and 90 by France. The figures are unfortunately missing for most Mediterranean countries.

The largest numbers of endangered forest-occurring bird species (> 20 endangered species) have been reported for Sweden, the Baltic countries and some countries in Central and Eastern Europe (i.e. Switzerland, Austria, Czech Republic, Slovakia, Slovenia and Albania).

The TBFRA-2000 data on the total number of bird species partly refer to the breeding and partly to the total number of bird species inclusive of migratory birds (see Chapter 6.1). It seems likely that the numbers provided for forest-occurring species mainly contain species breeding in the country, but the results should still be interpreted very carefully.

Figure 23. Number of forest-occurring bird species reported for different European countries

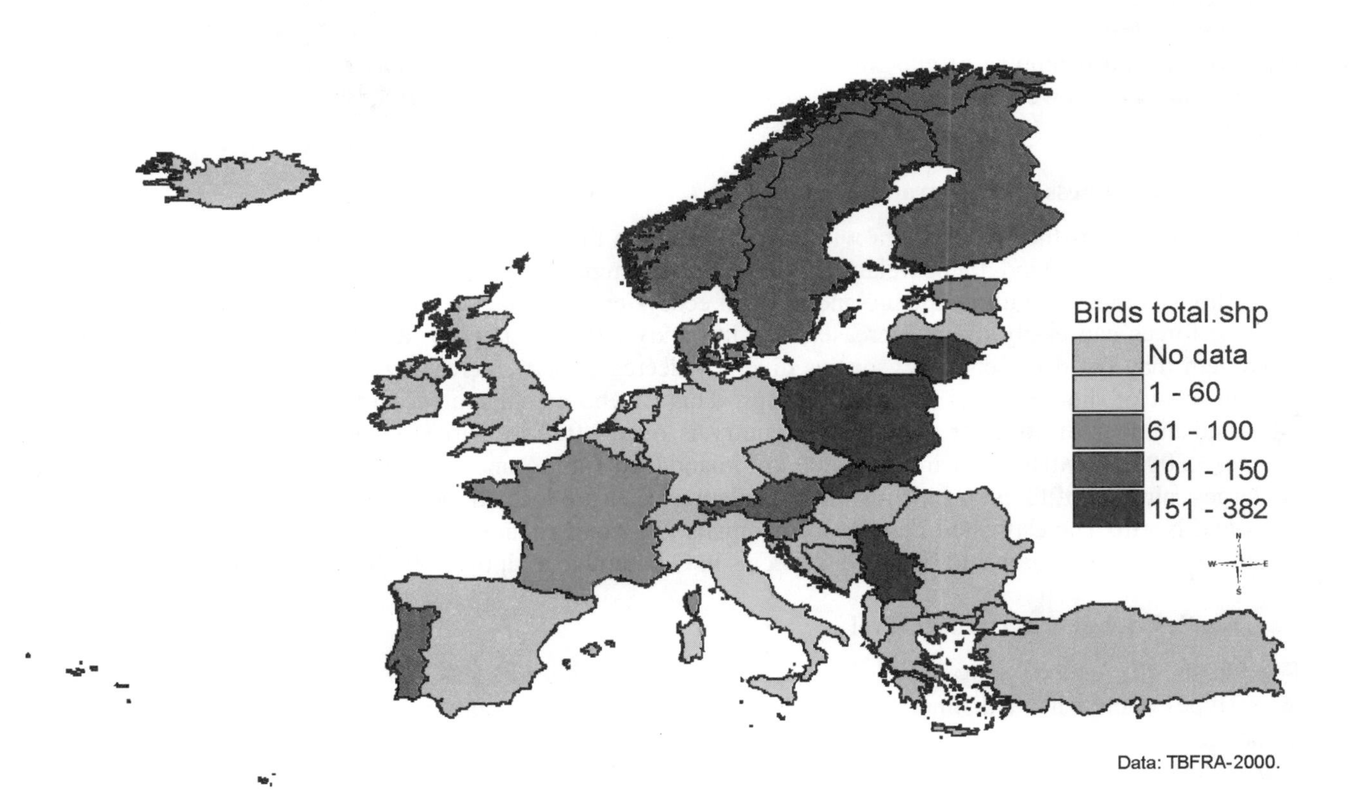

5.3 Number of Species per Unit Area

Globally, Europe does not count as a "biodiversity hotspot" in terms of high species richness. Myers et al. (2000) selected 25 biodiversity hotspots on the Earth based on species endemism and degree of threat. A selected area contains at least 0.5% or 1500 of the world's plants as endemics and should have lost at least 70% of the primary vegetation. Even if the hotspots only cover 1.4% of the land surface of the Earth, they contain 44% of vascular plant species and 35% of the mammal, bird, reptile and amphibian species on the Earth. The Mediterranean basin has fulfilled this biodiversity hotspot criteria as the only area in Europe (Myers et al. 2000). Caucasus is the closest area to the East.

Mediterranean biodiversity "hotspot"

According to some estimates, the Mediterranean forests have nearly twice as many tree species as other European forests (247 vs. 135, respectively) (Quézel et al. 1999). The number of native tree species reported in the TBFRA-2000 is about double in Mediterranean countries (France, Italy, Slovenia and Portugal) compared to Nordic countries (Finland, Norway and Sweden). The corresponding figures in Southern and Northern countries are 63 - 73 vs. 32 - 33, respectively. 5-10% of the total floristic species in the Mediterranean area are endemic (Barbero et al. 1990) and according to the IUCN 53% of the endemic floral species are currently threatened. Quézel et al. (1999) state that most southern Mediterranean forests are to some extent endangered. In the Eastern Mediterranean, especially in Turkey, the situation is somewhat better due to strict forest management activities. Lack of a full-scale survey on endangered forest ecosystems in the area, however, makes exact estimation difficult. Examples of endemic conifers in the Mediterranean basin are some firs, cypresses and pines (e.g. *Abies pinsapo, Abies marocana, Cupressus atlantica* and *Pinus nigra subsp. Dalamatica*) and examples of endemic deciduous species some oaks (*Quercus euboica, Quercus vulcanica, Quercus aucheri*) as well as *Liquidambat orientalis* in Turkey (e.g. Quézel 1998, Barbero et al. 1990, Akman et al. 1993).

Species richness and geographic location

Some areas such as the Artic tundra will never make it to the lists of biodiversity hotspots based on species richness, since they tend to be relatively simple ecosystems and low in species numbers and in trophic interactions compared with temperate and tropical ecosystems (Weider and Hobæk 2000). In these conditions, external forces can easily mean decreased stability and increased fragility, and the biodiversity native to the ecosystem may be lost. Generally, species diversity decreases with increasing latitude (e.g. Fischer 1961), at least in most of the terrestrial organism groups. This can be demonstrated even at the European scale: The estimated number of vascular plants in a country is over 9,000 in Turkey, around 5,500 in Italy, 4,500 in France, 3,200 in Germany and around 1,200 in Finland (see Table 8 and Figures 24 and 25). The countries with the largest number of forest-occurring vascular plant species are located more in the South and East than in the West and North. The estimated number of mammals in a country ranges from over 130 species in Turkey, to around 100 species in Italy and France, 90 species in Germany to a bit over 60 species in Finland (see Table 9).

Species richness at country level

The forests of a country may show a rich species diversity either because there is a large range of different forest types each with a distinct biota or because the forest types present are highly diverse (Kapos and Iremonger 1998). In Europe, the small countries are the most "species-rich" countries, if the total number of forest-occurring species, be it trees, other vascular plants, birds or mammals, is related to the unit area (i.e. species number divided by the area of forests and other wooded land in the country) (Figures 23 and 24). These countries also tend to have the largest number of endangered species per unit area.

The number of forest-occurring tree species does not generally increase as a function of the area of forests and other wooded land at the country level in Europe (Figure 26). This seems to apply for all the forest-occurring species of the taxonomic groups reported in the TBFRA-2000. The following reasons are likely to contribute to this phenomenon:

- According to island-biogeography theory (MacArthur and Wilson 1967) the species-area curves have generally a positive, non-linear relationship. It seems that when the percentage of suitable habitat in the area is higher than 30 % (Andren 1994) the species composition or population distribution can be predicted by analysing the reduction in the area of the suitable habitat (Haila 1983). When the suitable habitat coverage falls below 30% of the total land area, the division of the suitable area into patches and the spatial distribution of the patches start to affect the species. The responses are species-specific. European countries where the percentage of forests of the total land area is over 35% are Albania, Austria, Estonia, Finland, Latvia, Lichtenstein, Portugal, Slovakia, Slovenia, Sweden and the FYR of Macedonia (TBFRA-2000).

  In Europe the forest cover alone does not form the "suitable habitat" for forest-occurring species. Some of the species are related to old-growth forests, some to early succession stages and some to certain forest types. Most forest-occurring insect species live in scattered micro-habitats such as dead wood or rare host plants (Hanski 1999). Therefore direct conclusions cannot be made on the basis forest coverage. If we, however, think that in most European countries the proportion of forest from the total land area is around 1/3 or less it becomes obvious that the fragmentation of habitats into patches is bound to have an effect on the species in most countries. Patch area and isolation affect the population density (Hanski 1999). Density is expected to decline with increasing isolation, since isolation generally reduces immigration but affects emigration less. Small and isolated patches are most likely to be "empty".

- There is normally an asymptote for the species richness as a function of the size of an area in question (e.g. habitat, region and bio-geographical region) (Palmer and White 1994, Vanclay 1998). Since the small European countries show high species per forest area relationships, they seem to be large enough to support a certain number of species and reach this asymptote.

- The largest forest areas in Europe are found in the Nordic countries or in countries with considerable mountainous areas. These areas and the corresponding ecosystems, even if large in extent, naturally contain a smaller number of species than for instance a Mediterranean forest ecosystems.

It seems that country-level analysis on the biodiversity hotspots and species-rich areas in Europe does not make sense. Firstly, the small countries turn out to be the most species rich countries, independent of the taxonomic groups. This seems to be a symptom of wrongly selected scale, i.e. the small countries are already large enough to contain a large variety of forest occurring species and the number of species does not considerably increase when a certain threshold size of an area is reached. This seems to apply at least to the Central, Eastern and Northern European countries. The data for the most Mediterranean countries were unfortunately missing. A further affecting factor in this context is likely to be the historic background in the formation of countries. The administrational and political divisions of Europe as a continent are artificial and fail to represent ideal reporting units for an issue such as biodiversity.

European countries represent different bio-geographical zones so that the natural number of species is different in different countries (and in different bio-geographical regions). Most of the European countries are also so large that they contain considerable variation in forest types even within one country. Some localities may be especially rich in species, but after the data have been averaged for the whole country this is not clear anymore. Therefore it is more important to locate the valuable areas within each country, forest type and bio-geographical region – rather than to carry out evaluations at a country level.

Figure 24. Number of vascular plant species per unit area of forest and OWL (1000 ha) in a country

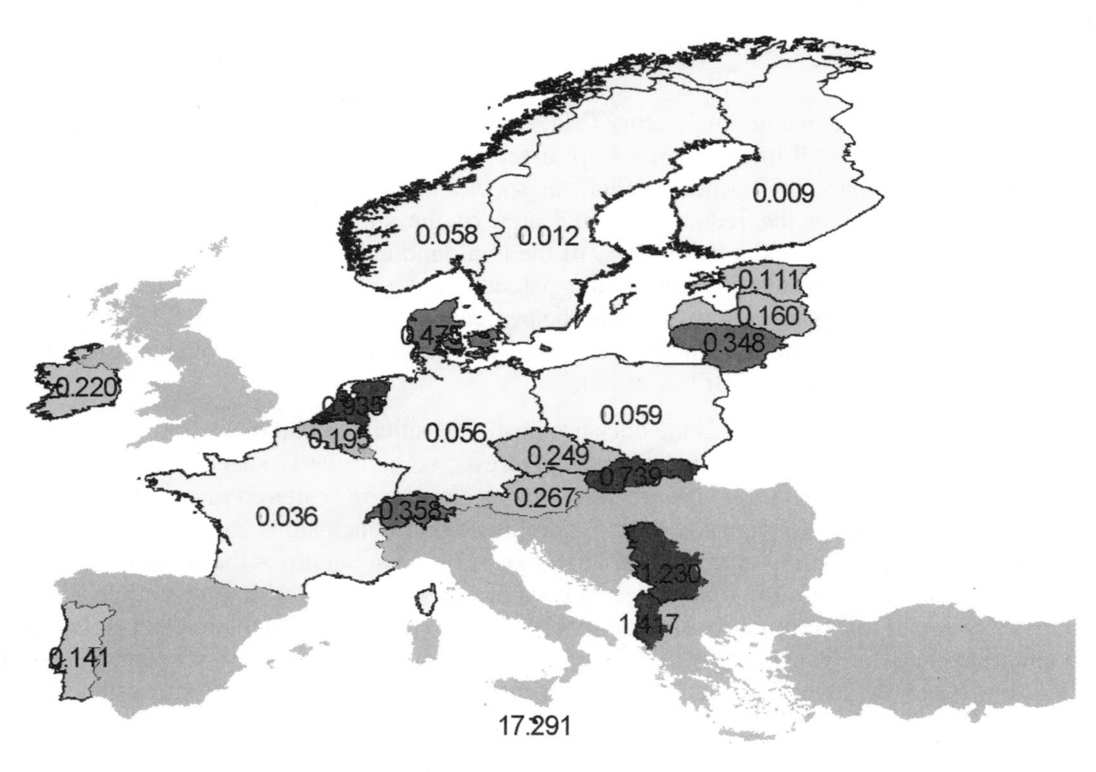

Figure 25. Number of forest-occurring tree species per unit area of forest and OWL (1000 ha) in a country

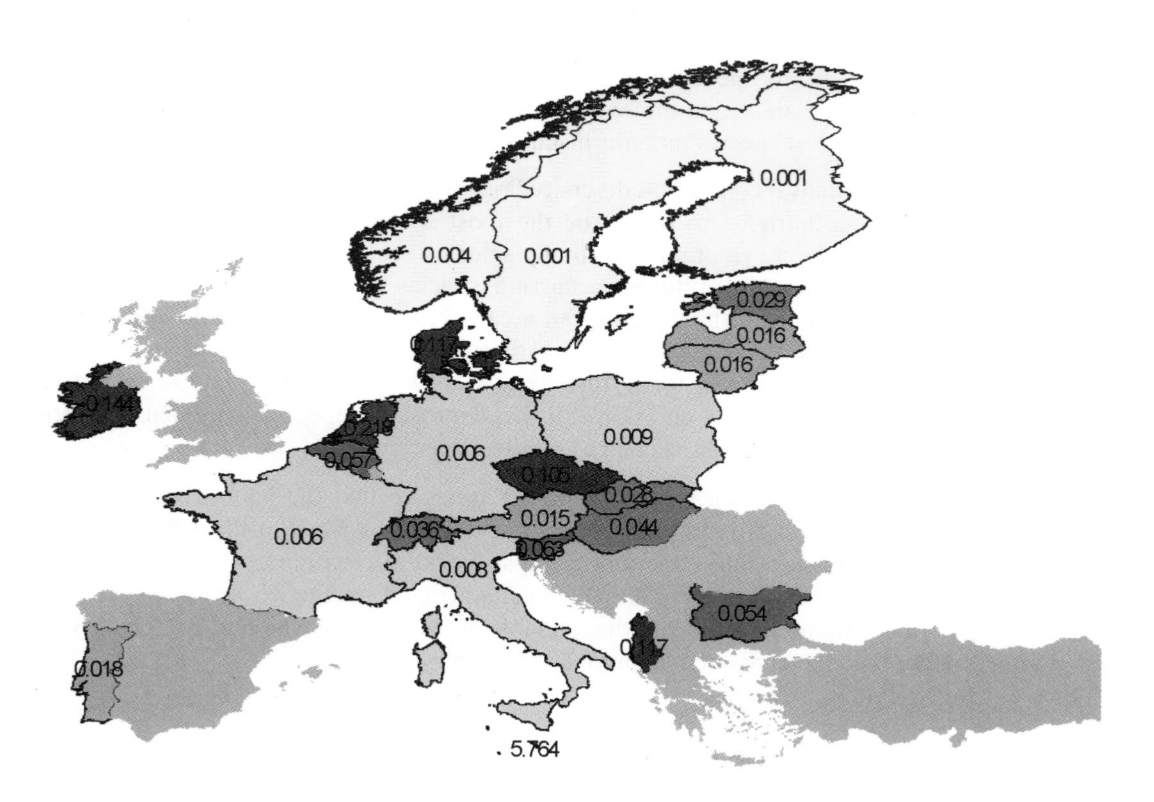

Figure 26. Number of forest-occurring tree species as a function of the area of forests and OWL. Each dot represents one country. Data: TBFRA-2000.

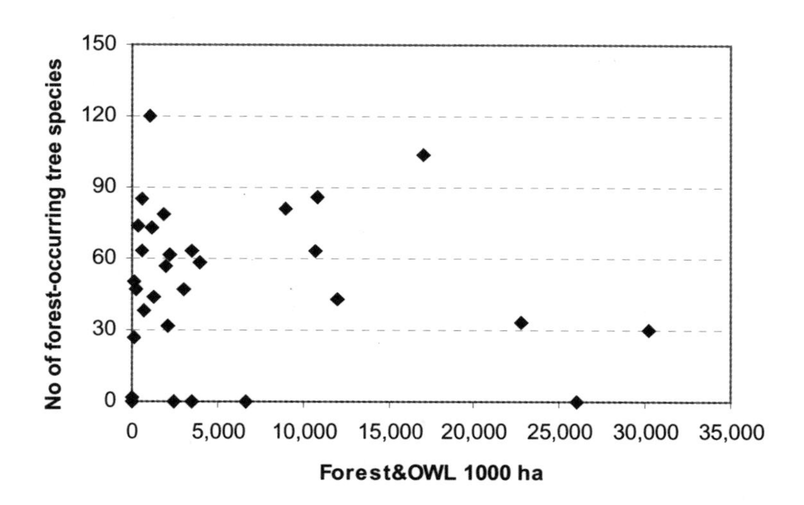

5.4 Introduced Species

Homogenisation of landscapes and introduction of exotic species are expected to turn the world's biota into more and more similar ecosystems (e.g. Niemelä 2000). Introduced species are also used in Europe for forest establishment (Figure 27). They are commonly used in transforming former agricultural lands into forests. It is, however, not clear to what extent introduced species are used in replacing harvested woodlands of native species. The extent of the current forest cover occupied by introduced, non-native or not "site-original" species is also not known. In total, around 200,000 ha of forest and other wooded land are annually created with introduced species in Europe. This corresponds roughly to over 20% of the total area created. The introduced species play a clear role in forest creation in the British Isles, South-eastern Europe and in Hungary and Sweden. Other countries where at least 20% of forests are annually created by using introduced species are Belgium, Denmark, Iceland and Israel.

Figure 27. Annual area of introduced species used in forest regeneration, extension and natural colonization. Data: TBFRA-2000.

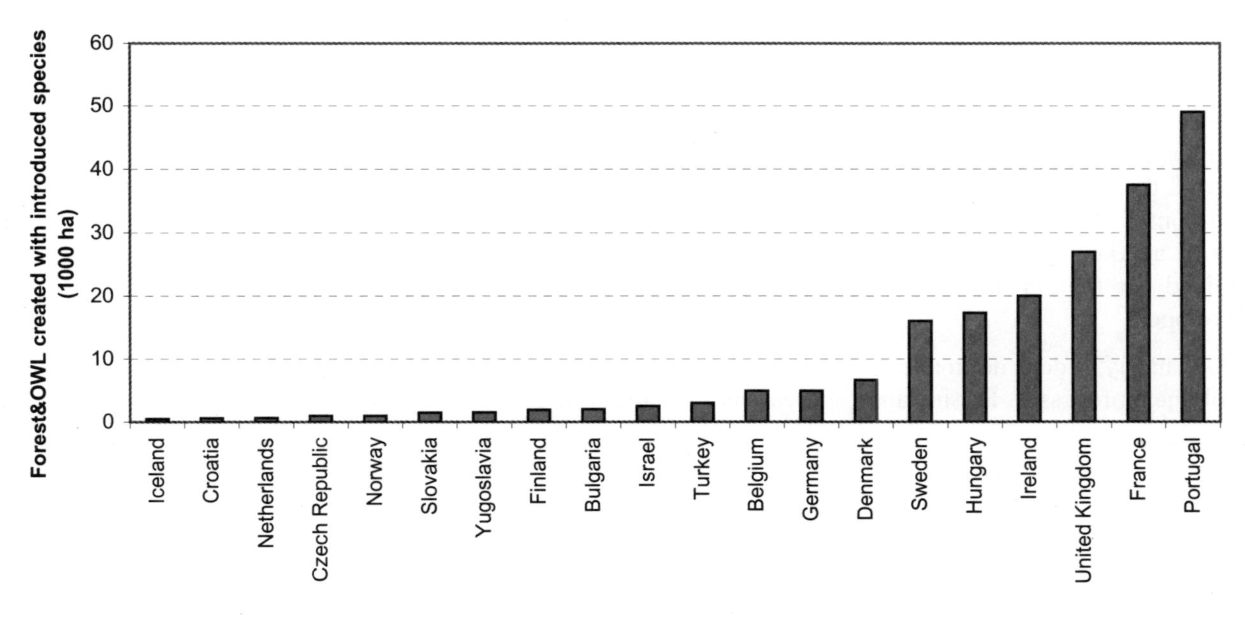

6. Functional Aspcest fo Forest Biodiversity in Europe

The role of 'function' as a component of biodiversity has remained more vague than composition and structure. According to one definition 'function' involves ecological and evolutionary processes and has an influence, on one hand, on the processes such as photosynthesis, nutrient cycling and population growth and, on the other hand, on system structure (Larsson et al. 2001). Concepts such as ecosystem functions, ecological processes, functional variety or functional relationships, which can be found in the different definitions of biodiversity[25], indicate that the variety of ecological processes belongs to the concept of biodiversity – at least according to current understanding. The differentiation between the processes or functions as such and their role in biodiversity seems to be incompletely addressed.

A series of ecological and evolutionary processes takes inherently place in ecosystems. Examples include photosynthesis, nutrient cycling, decomposition, forest fires, constant processing of the balance between different populations or the functioning of the carbon flux between the forests and the atmosphere. The outcome of these natural processes can be continuous or discrete, for example tree growth or falling of a decayed tree on forest ground. In case of discrete events we often talk of natural disturbances. Generally, natural processes result from the functioning of different organisms within an ecosystem. The state of an ecosystem in a particular time is the outcome of not only natural processes, but also of human influence. Human influence may change the landscape radically and quickly or indirectly and gradually over a long period of time, for instance in the form of climate change. Thus, effects of both natural processes and human influence may be either continuous or discrete.

The functional component of forest biodiversity is closely related to structure and composition. Structure is the framework in which the system functions and which restricts and guides the system's flows. Certain structures may not allow certain types of functions. For instance, in order to maintain a balance between species related to pioneer succession stages and old forests, sufficient amounts of both with an appropriate spatial distribution are needed. If no old forests are available, if they are too fragmented or located only in a small area, then the balance between the pioneer and old growth species is likely to be disturbed. Thus, by monitoring appropriate structural aspects certain premises on the functions can be made. For instance, average age class distributions (see Figure 10) serve this purpose.

Difficulties in the definition and monitoring of 'function' as a component of biodiversity

It is not trivial to link ecological and evolutionary processes, functioning of different organisms and anthropogenic influence in an ecosystem with forest biodiversity – or with the assessment and monitoring of forest biodiversity. Firstly, there is a danger of trying to encompass "everything" and ending up with an attempt to describe the whole ecosystem and its natural functions in general. Since there is neither for forest biodiversity nor for the functional component of biodiversity a clear single definition, this problem is not an easy one to put aside.

Secondly, there are approaches, which see the functional component merely as a product of structural and compositional aspects. For instance, in an ecosystem with a heterogeneous structure and a large number of species there is likely to be more interactions and functions between different organisms than in a structurally and compositionally simple forest. However, either there should be an understanding that forest biodiversity can be adequately described by using the structural and functional aspects alone or a distinct scope should be derived for the functional component. Currently, the former is not proven and the latter not adequately developed.

Thirdly, it does not make sense to determine the variety of functional component via the variety within the ecological processes. For instance, the variety of photosynthesis can most likely be increased by disturbing the

[25] Some examples of definitions: "Function involves ecological and evolutionary processes, including gene flow, disturbances and nutrient cycling" (Noss 1990). Functional factors at different levels of the hierarchy according to Noss (1990) are disturbances, land-use trends, land-use processes, inter-specific interaction, ecosystem processes, demographic processes, life histories and genetic processes. Larsson et al (2000) have included natural disturbance and human influence into the functional key factors of biodiversity as described in Table 1. They also state that functional diversity may refer either to the diversity of the ecological functions performed by different species, or to the diversity of species performing given ecological functions.

normal patterns of photosynthesis. This would yield an increased number of different photosynthetic and biomass growth patterns - as the results on forest growth patterns close to known pollution sources suggest (e.g. Pretzsch 1999b). In this respect, the most sensible interpretation brings one to the concept of ecosystem integrity (Angermeier and Karr 1994; De Leo and Levin 1997). So the functional component can be seen to contribute to the ecological integrity, to the ability of ecosystem to function and maintain itself and to the ability of the ecosystem components to generate and maintain forest biodiversity (also referred to by Larsson et al. 2001). Such definition, however, is not an easy one to synthesise in assessments and monitoring.

The fourth problematic aspect related to the monitoring of the functional component is that it includes both natural and anthropogenic and continuous and discrete events. Especially for monitoring purposes it seems to make sense to differentiate primarily between continuous and discrete processes and secondarily between natural and anthropogenic effect. The approaches and techniques to monitor continuous and discrete processes are somewhat different. However, it is not always easy to differentiate between natural and anthropogenic influences. For instance, human-induced global change contributes to the increased intensity and frequency of extreme weather events, which in turn cause an increase for example in forest fires and wind damage. Forest fires, furthermore, may be directly human-induced in some areas (e.g. due to tourism and political land-use conflicts).

Fifthly, the contribution of a single factor to forest biodiversity may be both positive and negative at the same time. For instance, the extent of burnt areas and succession stages related to them has become rare in temporal and boreal forests (e.g. Kouki 1994), whereas the temporal and spatial intensity of forest fires has increased in the Mediterranean countries. Therefore, it is difficult to set clear targets or even directions for some of the functional aspects, at least at a national level. Different forest types are characterised by different types of natural dynamics (e.g. Bengtsson et al. 2000). For instance, Larsson et al. (2001) recommend that monitoring of the majority of the functional factors should be undertaken at a landscape level and that reference states are urgently needed for different areas and forest types.

These arguments indicate that work is still needed prior to the establishment of concrete and comprehensive indicators on the functional aspect of forest biodiversity. Therefore, the description of this aspect is the most limited one in this paper. As in the previous chapters, the analysis framework utilizes the work done on key factors of European forest biodiversity carried out by Larsson et al. (2001). These functional key factors comprise the natural and anthropogenic influence and emphasise the monitoring of disturbances or discrete events (see Table 1). The national level data on the disturbances are well documented in the TBFRA-2000 Main Report. Thus, these data are complemented with notes on some central continuous processes, natural and man-made, in forest ecosystems.

6.1 Monitoring of Continuous Processes

Continuous processes related to forest ecosystems may include beside natural processes such as photosynthesis and nutrient cycling also anthropogenic influence such as pollution. As the functional component of forest biodiversity contributes to the ability of the forest ecosystem to function and maintain itself, such measures, which ensure that healthy levels of the natural processes are maintained and that possible changes are detected, are needed. Generally, there should be measures and measuring devices relevant to ecological phenomena, capable of providing continuous assessments and of differentiating between natural cycles and human-induced stress, sufficiently sensitive for early warning, geographically representative and realistic to be assessed in practise (Noss 1990).

Some monitoring methods and practices are already in use and can be seen to contribute to forest biodiversity assessments in that they are principally capable of revealing deviations from the natural patterns of ecosystem processes. These are especially the monitoring of forest growth and crown condition, with some limitations. Monitoring systems for both currently exist and are developed further in different instances in Europe. Information on forest growth primarily serves the forest planning at different temporal and spatial scales and data on crown condition are collected mainly for health monitoring.

Forest growth

Forest growth, i.e. the volume growth of trees as understood by foresters, is an aggregated, robust, stable and unspecific indicator for describing the condition of forest ecosystems (e.g. Pretzsch 1999b). Forest growth results from several processes, mainly photosynthesis and nutrient and water cycling. Small and temporal disturbances are normally buffered in the complex growth processes, and, if changes in forest growth occur, they are likely to indicate fundamental changes in growth behaviour and processes. This, however, also means that healthy growth behaviour does not guarantee that "everything is fine". Forest growth is an unspecific indicator because it does not necessarily reveal the reasons for the changes.

From a practical point of view forest growth is an appropriate indicator: Established monitoring systems and standards exist, they cover large areas, are relatively cost efficient and the growth curves accumulated over a long period of time allow the detection of changes. This has been clearly demonstrated in the detection of the upward trend in forest growth in Europe over the past decades (Spiecker et al. 1996). The accelerating forest growth is likely to be related to a complex mixture of a rise in temperature and prolonged period of seasonal growth, nutrient inputs, local site conditions and biotic stress factors (e.g. Pretzsch 1999b).

All in all, the relative net annual increment (as a percentage of the growing stock) is lower in Eastern than in Western Europe (Figure 28). This is likely to be related to the intensively practised silviculture in Western Europe. In absolute terms (m^3/ha) the net annual increment is the largest in Central and Western Europe (above 5 m^3/ha o.b.) and somewhat lower in Northern Europe and Mediterranean countries (Figure 29). The main factor limiting the growth in Northern Europe is the length of the growing season and in Southern Europe the availability of water. This level of spatial and temporal detail, however, only allows very rough judgements of the current situation.

Crown condition

Crown condition, i.e. changes in defoliation and discolouration, is also an aggregated and unspecific indicator of forest condition. Crown condition is a result of several processes, and defoliation and discolouration caused by pollution cannot be separated from changes caused by other agents or environmental processes (e.g. natural characteristics and variation, age, site conditions...). Currently especially weather extremes, air pollutants, changed soil condition, insect attacks and fungal infestations influence the condition of European forests (EC and UNECE 1999a). Influence of individual factors also depends on tree species, forest type and geographical location, for instance.

Crown condition in Europe has been monitored for 14 years within the ICP-forests[26]. Currently, a network of about 5,700 plots provides crown condition data. Recent monitoring results indicate that the overall deterioration of crown condition has slowed down (EC and UNECE 1999a). Still, almost a quarter of the assessed trees are classified as damaged.

Indirect anthropogenic influence

Indirect human impact is caused mainly through policy-making and effects such as human-induced climate change. Indirect effects tend to become apparent gradually over a long period of time, and the establishment of direct causal relationships may be difficult due to the complex net of interrelated factors. Some examples of political influence are the current favouring of deciduous trees in forest regeneration or the development of management methods for continuous cover forests. Also the favouring of the afforestation of agricultural lands in Europe and the fact that the Common Agricultural Policy of the European Union prior to the 1990's encouraged the ploughing up of grasslands, clearing hedgerows and increasing field size serve as example. This intensified farming led to the general loss of diversity in rural landscapes. This kind of dangers have been recognised and, for example, the Biodiversity Action Plans of the European Commission emphasise the importance to integrate biodiversity needs into the development and implementation of relevant sectoral policies (EC 2001).

---

[26] International co-operative programme on assessment and monitoring of air pollution effects of forests (ICP Forests) of the United Nations Economic Commission for Europe (UN/ECE) and under the Scheme on the protection of forests against atmospheric pollution of the European Union (EU) (EC and UNECE 1999a,b, see also Chapter V of the TBFRA-2000 Main Report).

Figure 28. Net annual increment of forests available for wood supply as percentage of growing stock

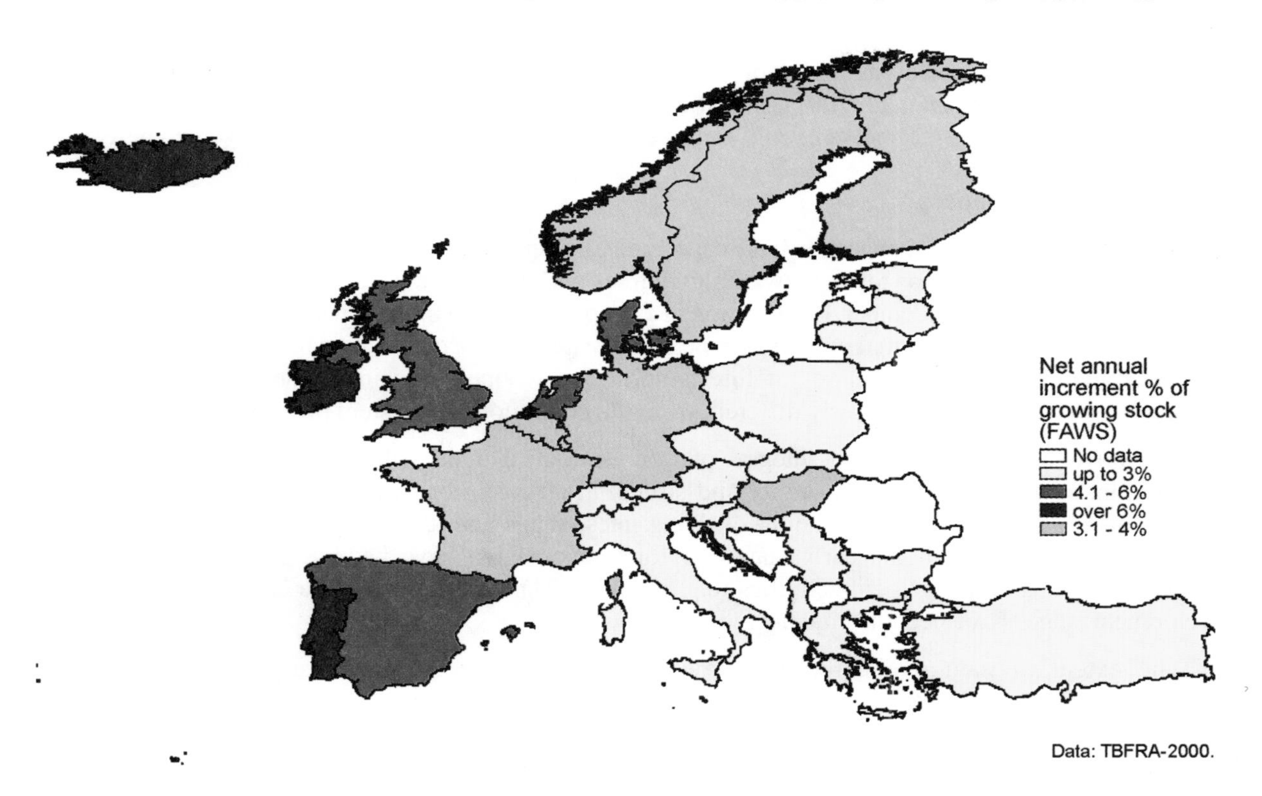

Figure 29. Net annual increment of forests available for wood supply (FAWS), m^3 per hectare

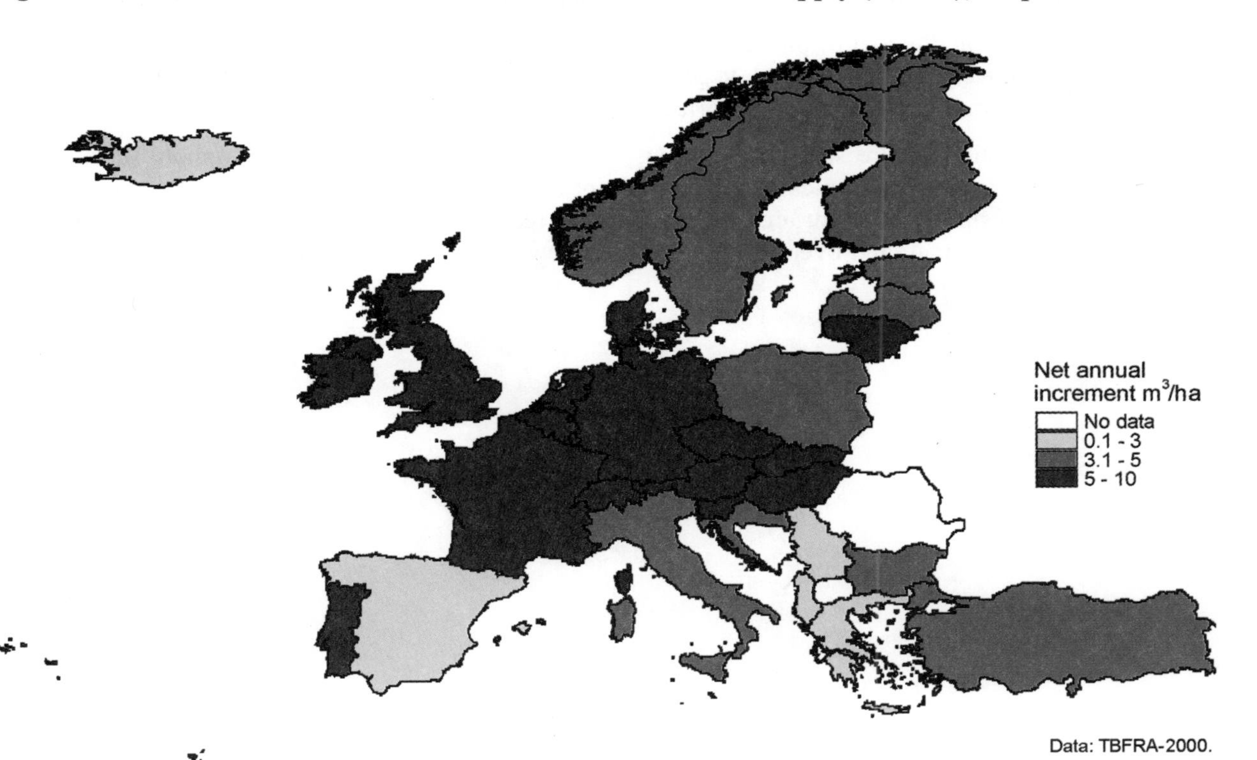

6.2 Natural Disturbances and Discrete Anthropogenic Influence

The functional key factors of forest biodiversity include the natural influence of fires, wind and snow and biological disturbance as well as the anthropogenic influences of forestry, agriculture and grazing, other land uses and pollution (Larsson et al. 2001). Statistical data on these aspects are well provided by the TBFRA-2000 Main Report (see Table 3). Therefore, only some selected aspects are discussed and presented here.

Natural disturbances

Natural disturbances may change the landscape radically and quickly, and the scale of the disturbance may range from large burned areas to single windblown trees within a forest. The same applies to human influence[27]. For instance, due to the December 1999 storms in Europe the total windblown volume was estimated to be 165 million m^3, or 43% of the normal annual European harvest and 20% of the annual growth[28]. Furthermore, forest fires burn about 0.5 million hectares of forests and other wooded land in southern Europe every year. The nature of this kind of events is clearly different from the continuous ecosystem processes.

Natural disturbances form an integral part of ecological and evolutionary processes and they initiate regeneration, succession, habitat diversity and structural change. For instance, the extent of burnt areas with large amounts of dead wood, the amount of dying trees, snags and logs lying on the ground and a deciduous component associated with the early stages of forest succession have become rare in temporal and boreal forests due to forest management activities (e.g. Kouki 1994). Subsequently, the communities related to these development stages have become endangered.

The extent of natural disturbances may be expanded due to human activities. Forest fires in the Mediterranean area and large wind damages in Central European plantations serve as an example. To some degree the disturbances are an integral part of ecosystem dynamics, but may become over-extended due to unfavourable human impact. Vegetation fires also significantly contribute to the release of carbon dioxide into the atmosphere both directly through the emissions of gases and aerosols from the fires and indirectly through the impact of fire activity on the forest ecosystems and on their capacity to act as carbon pools.

Beside the origin and extent of the disturbance the actions taken after it are important. After a large-scale disturbance the recovery of the natural vegetation with subsequent succession phases may follow – or transformation into shrub-land, plantations, agricultural area or even urbanization.

Periodic publication of fire statistics indicates long-term trends[29]. Additionally, remote sensing offers a good alternative for a gross and quick estimation of fire activity for fairly large regions. Recent studies have demonstrated the feasibility of burnt area estimation with remotely sensed imagery, at spatial scales ranging from regional to continental, and over time periods of a few days to multiple years (e.g. Barbosa et al. 1998; Barbosa et al. 1999).

---

[27] Human influence may also be introduced indirectly and gradually over a long period of time, for instance in the form of climate change. This has been referred to in the connection of the monitoring of continuous processes (Chapter 6.1).

[28] Press Release ECE/TIM//00/2 Geneva, 18 January 2000 (FAO/UNECE Europe).

[29] Fire statistics are collected and evaluated for instance by the UNECE Trade Division, Timber Section, Geneva.

Direct anthropogenic influence

European forests develop in close contact with the surrounding society and reflect the interactions between the people, economy and ecology. Forestry professionals and forest owners have a direct impact on the forests through forest management activities. The general public affects the forests directly for instance through recreation, tourism and by picking berries and mushrooms. The direct involvement of the general public in forest management is also increasing so that the different opinions are taken into account by using participative planning methods.

In total, 88% of forests and 63% of other wooded land are managed "in accordance with a formal or informal plan applied regularly over a sufficiently long period..." (TBFRA-2000). 15 countries out of 38 report that the complete area of forests and other wooded land is managed (Austria, Belgium, Cyprus, Czech Republic, Denmark, Germany, Hungary, Latvia, Liechtenstein, Malta, the Netherlands, Poland, Romania, Sweden and Turkey)[30]. The majority of forests are managed as even-aged high forests (Figure 30). Uneven-aged forests exist in the long run either because a stand has been left for free development or because it is managed by using a selection cutting system. Generally, it seems that the selection cutting management of forests is increasing especially in Central Europe (e.g. Gadow and Puumalainen 1998), even if there are currently little data on the actual area managed by using selection cutting management. Only a minor proportion of forests and other wooded land are under limited human impact (see Chapter 5).

Figure 30. Main forest management methods applied in Europe as percentage from the forest area available for wood supply in Europe. Data: TBFRA-2000.

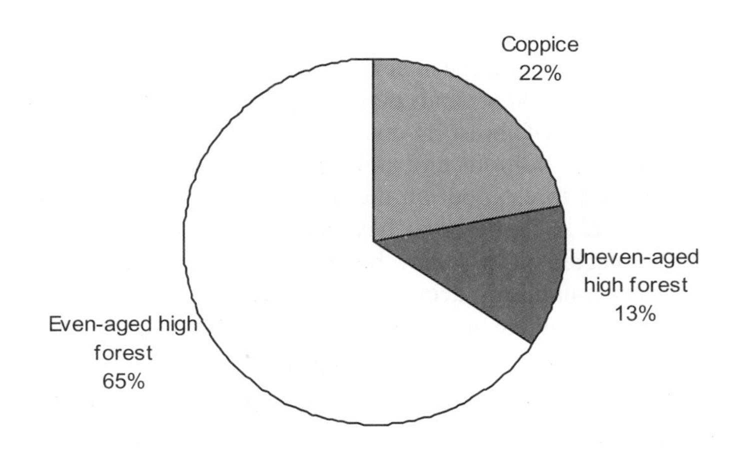

[30] Some Mediterranean countries (Portugal, Greece and Italy) as well as Luxembourg, Albania and Iceland report the proportion of managed forests and other wooded land to be less than 40% of the area.

7. Discussion and Conclusions

This work arose from the current deficiencies and difficulties related to international reporting, large-scale assessments and long-term monitoring of forest biodiversity – as well as from the interest in the variety of European forests. A pragmatic approach was taken so that despite of missing universal definition of biodiversity, non-existing "final" list of biodiversity indicators and lacking "perfect" data, an assessment of European forest biodiversity has been made.

Generally, this paper is not there to state if the forest biodiversity in Europe is good, bad, adequate or only 70% of what it should be. Such evaluations are difficult to make, since they depend on the diverging values and partly contradictory goals of the society. The understanding of the current situation, past developments and of the functioning of the ecosystems merely assists in setting priorities and goals for the future policies and actions. In order to maintain appropriate levels of biodiversity and ecological integrity we should continuously improve this understanding at different levels of the hierarchy and assist policy making in the follow-up of the developments and in the improvement of selected criteria and indicators, monitoring, and in the setting up of new targets as new knowledge becomes available.

This discussion paper aimed to improve the understanding of the current diversity of forests in Europe and further to detect strengths, improvements and alternatives for future assessments. The main body of the work has concentrated on the comprehensive description of current state by using the framework of structural, compositional and functional aspects of forest biodiversity. These results are synthesised in the summary, and not repeated here. Instead, attention is given to the implications of the detected variety and strengths and difficulties of the assessment and monitoring.

Variety in European Forests

Dealing with forest biodiversity "on a European scale means dealing with great regional diversity in ecosystem character and social systems, forest history and, ownership structure" (Andersson et al. 2000). Therefore, it is not surprising that different aspects of forest diversity are accentuated in different ways in different European countries, due to variable natural conditions and anthropogenic effects. A single country may have on the one hand a large proportion of mixed forests, but on the other hand mainly even-aged forests. In another country these factors may be the opposite. A country may have a relatively large proportion of strictly protected forests, whereas the total area of protected forests may be moderate compared with another country. The species richness in terrestrial ecosystems naturally decreases with increasing latitude, and the most important habitats for biodiversity protection are not visible at national level assessments. Silvicultural regimes vary according to the forestry tradition but also according to different forest types.

The large variety even at such a low level of resolution like the national level emphasizes that one or only a few indicators can rarely capture the true variety in the forest structure in Europe and that the attempts to assign one comparable value for the forest biodiversity are very limited and un-informative. It is difficult to do justice for the true diversity and natural integrity in a single indicator, as the variety is expressed in different parameters in different localities.

Furthermore, the ecological understanding of the contribution and "value" of each factor to the overall biodiversity is poor, and always depends on the natural conditions of the locality and forest type. "To understand diversity patterns a necessary first step is to place these patterns in appropriate perspective. Diversity values are exceptionally difficult to interpret when taken out of the context, and little justification exists for their publication if such perspective is not provided" (Peet 1978). Therefore, the application of a number of selected indicators and their interpretation in an appropriate context should be further encouraged in forest biodiversity assessment - even if the approach may seem too diverse for someone seeking a single figure or a simple answer.

Framework of indicators

Criteria and indicators for describing forest biodiversity are in constant development as the understanding of

- Ecological processes,
- Interactions between the composition, structure and function,
- Linkages between ecological, economical and social development

refines the selection of the most appropriate ones. There are numerous suggestions for appropriate indicators and for their desired characteristics. Altogether the selected measures should be purposeful, technically and scientifically feasible as well as ecologically relevant. Quantitative measures are needed to objectively and reliably describe current situation, state the aims and monitor the progress in the future. Qualitative measures may especially help to understand some of the issues involved – but are difficult to monitor or compare.

In this paper, a fairly pragmatic approach was used to select the indicators. A framework for comprehensive description and, further, a set of measures for which existing, credible and comparable data were at least to some extent available, were sought after. Mainly quantitative measures have been used. The framework of structural, compositional and functional key factors of forest biodiversity (Larsson et al. 2001) allowed a comprehensive and structured analysis of the variability of forests. Within the framework different aspects of biodiversity are taken into account so that the variety of European forests is captured.

Generally, population-species level attains the main focus of the biodiversity monitoring. This may be because the number of species is easy to define and understand, also for the common public, and easy to maintain attraction. However, the definition(s) of biodiversity comprise more than just the species richness and the consideration of species richness alone is likely to cause confusion and may be misused. Species richness should always be contrasted with the local natural conditions. Boreal forests are poor in species if they are compared with some Mediterranean forests – and Mediterranean forests are poor in species if they are contrasted with tropical rainforests. Some areas such as arctic tundra will never make it to the lists of biodiversity hotspots – but that should not mean that these ecosystems are not valuable. Furthermore, we can enhance species richness by increasing the number of exotic species in region, by counting species located in parks and arboreta and not only those in natural conditions, by observing a large area instead of a small one and through the detection of new species.

The estimates for the number of species (e.g. single species, genera, families) in a country considerably vary from different sources. The estimates may be based on different geographical area and species grouping. Breeding or migratory species may be count, new species are detected even in Europe all the time, and the intensity of the species counting varies. To detect a species one needs to be in the right place, at the right time and use the right method. Still, rare species are seldom sampled and some species such as fungi do not fruit every year and are therefore difficult to detect. Based on the comparisons carried out in this work (Chapter 5.1), it seems that the species count in a country provide a rough overview, but cannot credibly detect changes in species number and cannot be used for proper monitoring of biodiversity. Monitoring should be sensible enough to detect changes in the species composition before some of the species become extinct. Thus, on one hand holistic biodiversity monitoring and, on the other hand, early detection of changes require additional approaches and monitoring at landscape and stand level.

The number of endangered species also has some limitations, especially in international comparisons. A species may be threatened in a particular country because it is at the extreme range of its natural habitat – and abundant in Europe or in neighbouring countries. This is because the administrative borders do not necessarily coincide with sensible reporting units for natural phenomena. Furthermore, the risk of extinction is assessed differently and the use of diverging monitoring accuracies makes comparisons difficult. Varying classification schemes for threatened species are applied, despite of international standards (IUCN, WCMC) and national Red Data Books. Countries may record additional classes such as "near threatened". In Sweden, for instance, this doubles the number of threatened species in many taxonomic groups.

The criticism on the species count does not imply that it is a useless measure. However, considering the broad attention the species numbers gain, it seems appropriate to underline the possible difficulties with its interpretation and application as monitoring tool. Due to these difficulties the monitoring of biodiversity commonly uses not only the distribution and abundance of organisms, but also their associations with the physical environment (see also Niemelä 2000). The structural and functional key aspects provide one possible set of tools for this. For instance, the changes in the spatial structures of landscapes (e.g. fragmentation) provide an early indication of possible dangers for species and populations.

The components of structure, composition and function are highly interlinked with each other. For instance fire (functional factor) affects the species composition (compositional factor) and the age structure (structural factor), which again affect the susceptibility for certain disturbances (functional factor). As a further example, trees and stand structure affect birds or insects. Due to these inter-links and the fact that structures are often easier to quantify and monitor than the species or functions itself, structural diversity is a useful indicator not only on its own shake but also to provide indication of the other components.

The scope of the functional aspects of forest biodiversity is somewhat unclear. Improvements are particularly needed in separating and linking the appropriate aspects of continuous processes to forest biodiversity monitoring so that, on one hand, a distinct scope with respect to biodiversity is found and, on the other hand, not "everything" is included in the monitoring. These experiences only apply to a national level analysis. The value of the framework for landscape and stand level monitoring and applications needs to be tested in further studies.

Data for assessing and monitoring forest biodiversity

There is a competition in the production of (biodiversity-related) material for decision-making, and probably there is even more competition in the determination of the criteria and indicators. In these days of rapid communication it is possible to electronically publish information in no time. This may serve to save acutely threatened areas or improve the integration of the biodiversity point of view in conflict solving. However, the content and credibility of the presented information should be considered with care.

Generally, to make and keep biodiversity assessments and monitoring practicable, "everything" should not be included in them. Especially, there is little point in planning parallel monitoring schemes for different purposes. Therefore, the best option for biodiversity monitoring is to cooperate with the traditionally existing schemes and improve them to better take the biodiversity issues into account.

Compilation of existing data for biodiversity assessments generally offers a large scope for bias and confusion (e.g. Vanclay 1998). The original field data have arrived via several hands and heads, desks, computers and print-outs before they are compiled in this paper. The data have been collected in different countries according to different procedures, variables, definitions and standards during different time periods. The originally collected data have then been adjusted to fit the international definitions. Even the interpretation of the traditional forestry variables in such international efforts needs special care - and more so when we are interested in the estimation of change or in the biodiversity-related variables.

The cautiousness required to interpret the results has been underlined in the TBFRA-2000 Main Report. The same cautiousness applies for the results in this study. The quality of the data may have suffered especially due to

- Questions that required greater interpretation than the traditional forestry variables, which have been collected for a long time,
- Less complete and in some cases more ambiguous responses than those relating to more traditional issues,
- Misinterpretation of questions, and
- Disagreements with respect to interpretation of some information, particularly with regard to what constitutes a protected area.

There will most likely never be "the perfect data" available for forest biodiversity assessment and monitoring in Europe. Thus, this discussion paper tries to make the best out of the existing data, understanding their limitations. Generally, the TBFRA-2000 is a serious attempt to provide comparable data in an international assessment. On one hand, one should be aware of the shortcomings, but on the other hand, one should recognise the value of the data.

The TBFRA-2000 process provides data for describing almost all of the aspects required by the analysis framework. TBFRA-2000 directly provides data on eight of the twenty quantitative Pan-European indicators on sustainable forest management, five further indicators are covered partly or through data modifications and there is no information related to seven of the quantitative indicators. Further, information covering almost all of the structural, compositional and functional key factors – at least to some extent – is provided. Care is especially needed when interpreting the species numbers and protected areas. To some degree, different interpretations may affect especially the data on plantations, mixed forests and uneven-aged forests. In the work, the TBFRA-2000 data are also complemented with other data, studies and research results for validation, additional information and comparisons.

Scale issue in biodiversity assessment and monitoring

The interest in biodiversity provides an opportunity to address environmental problems holistically. "No single level of organisation (e.g. gene, population, community) is fundamental, and different levels of resolution are appropriate for different questions. Big questions require answers from several scales." (Noss 1990). National level reporting is appropriate for a broad over-view, for fulfilling international agreements and for comparing the effects of different policies carried out in different countries. National level reporting, however, is not an early-warning system. Before a species has become extinct at national level severe changes in the forest structure and species composition may have occurred – even a long time before (Hanski 2000). Providing that appropriate monitoring systems at a finer scale do exist, the changes may be detected in time for corrective measures.

National level reporting should be complemented with monitoring at a finer spatial and temporal scale (e.g. Dudley and Jeanrenaud 1998, McCormick and Folving 1998). This is necessary for early change detection, for detection of changes, which may not be visible at country level but only in certain ecosystems and communities, for setting preferences for protection areas and for the more detailed planning at a landscape and stand level. Especially by monitoring changes in the structural aspects dangers for habitat destruction and species extinction may be detected in time. This applies both for landscape and stand level. Even if the relationships between the structural, compositional and functional aspects are not clearly understood, it should not prevent one from making the best out of the current knowledge.

Managing and monitoring for forest biodiversity protection

Two complimentary approaches are used in practice to maintain and improve forest biodiversity, namely the improvement of the "quality" of managed forests and the network of protected areas outside normal forest operations. In managed forests, silvicultural activities are adapted to better mimic natural processes and retain valuable habitats. This nature-oriented silviculture is currently the main trend in European forestry. Generally it refers to somewhat less intensive management methods favouring retention trees, decaying wood, small-scale harvesting and protection of small key-biotopes (e.g. wetlands, river and lake boundaries) to improve the general quality of forests. The stand establishment with "natural" tree species and species mixtures, which especially in Central Europe often means favouring deciduous species, is also commonly cited.

The success of the policies towards nature-oriented silviculture should be seen within a few years in the values of the biodiversity indicators. The proportion of mixed forests as well as the amount of decayed wood should be increasing. The follow-up of the area under different management regimes should indicate an increase of uneven-aged forests. It is worthwhile to consider the separation of the area managed with nature-oriented silviculture from the other management methods – even though the definition problems within Europe will certainly be large.

The improvement of the management of production forests, however, may not reach the most specialised and also the most threatened species. Maintenance of large-scale biodiversity in production forests needs to be complemented with strictly protected areas. Core areas of natural forests are needed to retain some species – or to let them reappear and recover (Hanski 2000). Further, natural forests are crucial for the protection of natural processes and species related to these processes, for understanding the ecological principals, and as model for nature-oriented silviculture (Parviainen et al. 2000). There is also some indication that the total effort to improve and conserve forest quality should be directed into certain areas rather than distribute the same effort evenly but rather weakly throughout the landscape (Hanski 2000).

There are hardly any truly natural forests remaining in Europe, which is also confirmed in the data of this study. Due to the dense population forests have been subjected to influence of humans widely and for a long time, and any future solutions also need to consider the interactions between the man and the nature. The proportion of strictly protected forests is relatively small, especially in Central and Southern Europe, whereas the extent of other protection categories is considerable. This reflects the need to search and improve joint solutions.

Within the management and monitoring of protected areas in Europe further work is needed in clarifying the content and extent of all the 90 protection categories found in the area. This enables an improved assessment of the current state and of the areas, in which further developments are needed. There are also European level initiatives to establish representative networks of protected areas, such as NATURA and Ramsar wetland areas. These should be evaluated for their relevance with respect to forest ecosystems. The extent to which forest ecosystems are involved in these protection networks, their spatial distribution and how representative the areas are with respect to European forest types are currently not known.

Human-nature interactions and forest biodiversity

The holistic approach of assessing environmental problems includes no only the monitoring at several scales and by using a variety of different parameters but also an integrated analysis of ecological, social and economic phenomena. The history of the forests, landscape, people and their interactions is rich and complex in Europe. When the dependencies between the socio-economical development and diversity are studied, one deals with the fundamental – and not with the intermediate – causes affecting the biodiversity. Only an integrated approach in understanding the complex interdependencies between the socio-economic development and natural component would allow a proper understanding of the processes underlying the depletion of biodiversity.

The loss of biodiversity is often attributed to human influence such as logging, clearance of forest areas for other land use forms and pollution. These, however, are just intermediate reasons for the loss of biodiversity. "Fundamental causes are those underlying the proximate causes – they are basically socio-economic and institutional, and most of them are caused by the following factors: population growth, economic growth, technological change and institutional and policy arrangements" (Solberg 1998).

In Europe, the human impact has been important in shaping the landscape at least for the past 4000-5000 years (e.g. Bengtsson et al. 2000). Land-use patterns have evolved mainly around two factors: the type and accessibility of natural resources and the dynamics of demographic processes. These two are strongly interlinked. There is and has been a mixture of factors affecting biodiversity, and the trends for instance in agriculture, industrialisation and demography have been different in different countries. Only an integrated approach in understanding the complex interdependencies between the socio-economic development and the natural component will allow a proper understanding and prognosis of the processes underlying the depletion of biodiversity.

8. References

Akman, Y., Quézel, P., Ketenoglu, O. & Kurt, I. 1993. Analyse syntaxonomique des forêts de *Liquidamber orientalis* en Turquie. Ecologica mediterranea 19: 49-57.

Ammer, U. 2000. Gedanken zur Nachhaltigkeit, dem Zauberwort unserer Zeit. In: Landnutzungsplanung und Naturschutz. Aktuelle Forschungsberichte. Festschrift zur Emeritierung von Professor Dr. Ulrich Ammer. Wissenschaft und Technik Verlag, Berlin. P: 1-21.

Andersson, F.A., Feger, K.-H., Hüttl, R.F., Kräuchi, N., Mattsson, L, Sallnäs, O. & Sjöberg, K. 2000. Forest ecosystem research – priorities for Europe. Forest Ecology and Management 132: 111-119.

Anders, S. & Hofmann, G. 1997. Vielfalt in der Vegetation von Wäldern und Forsten. Summary: Diversity in the vegetation of natural and managed forests. Schriftenreihe des BML "Angewandte Wissenschaft", Band 465 "Biologische Vielfalt in Ökosystemen": 94 - 108.

Andrén, H. 1994. Effects of habitat fragmentation on birds and mammals in landscapes with different proportions of suitable habitat: A review. Oikos 71: 355-366.

Angermeier, P.L. & Karr, J.R. 1994. Biological integrity versus biological diversity as policy directives. BioScience 44: 690-697.

Anon. 1992. Iceland: National Report to UNCED. Ministry of Environment, Iceland. Reykjavik.

Anon. 1995. Materials on the situation of biodiversity in Germany. Federal Agency for Nature Conservation. Bonn. 112 pp.

Anon. 1997a. First Austrian national report on the Convention on Biological Diversity. Federal Ministry for the Environment, Youth and Family. Vienna. 63 pp.

Anon. 1997b. French report on the convention on biological diversity. Implementation of decision II/17 taken at the second Conference of the Parties to the Convention on Biological Diversity. The French Republic, French Ministry of Foreign Affairs and Ministry for Spatial Planning and Environment. Paris. 84 pp.

Anon. 1997c. First national report to the conference of the parties to the convention on biological diversity. Ministry of Environmental Protection, Natural Resources and Forestry of the Republic of Poland, National Foundation for Environmental Protection and UNEP. Warsaw. 58 pp.

Anon. 1997d. Convention on biological diversity – National report of the Republic of Slovenia. Ministry of the Environment and Physical Planning, State Authority for Nature Conservation. Ljubljana. 53 pp.

Anon. 1997e. Inform e español. Convenio sobre la Diversidad Biológica. Ministerio de Medio Ambiente. Mardrid. 51 pp.

Anon. 1998a. Biodiversity in Albania – Report on national situation of biodiversity in Albania. Nature Protection Directory, National Environmental Agency. Tirana.

Anon. 1998b. National report for the biological diversity conservation in Bulgaria. Ministry of Environment and Water. Sofia. 35 pp.

Anon. 1998c. Federal Government report under the Convention on Biological Diversity. Federal Ministry for the Environment, Nature Conservation and Nuclear Safety in Germany. Neusser Druckerei und Verlag GmbH, Neuss. 136 pp.

Anon. 1998d. Hungary: First national report for the implementation of the convention on biological diversity. UNEP & Ministry for Environment and Regional Policy, Office for Nature Conservation. Budapest. 64 pp.

Anon. 1998e. Biodiversity Conservation and Action Plan. Lithuanian Ministry of Environment. Vilnius. 108 pp.

Anon. 1998f. First Portuguese national report to be submitted to the conference of the parties to the convention on biological diversity. Ministry of Environment, the Nature Conservation Institute. Lisbon, Portugal. 117pp.

Anon. 1998g. Romania. National report for the biological diversity in Romania. Romanian Ministry of Waters, Forests and Environmental Protection. Bucharest. 34 pp.

Anon. 1998h. Turkey – First national report on the Convention on Biological Diversity. Ministry of Environment. Ankara. 38 pp.

Anon. 1998i. The United Kingdom national report on biological diversity. London HMSO. 56 pp.

Anon. 1998j. First Belgian national report on the Convention on Biological Diversity. Belgian Clearing House Mechanism. Brussels. 73 pp.

Anon. 1998k. National report on status and conservation of biological diversity in Slovakia. Ministry of the Environment of the Slovak Republic. Bratislava. 30 pp.

Anon. 1999a. National biodiversity and landscape strategy for Croatia. Narodne novine ("Official Journal) 81 (3.8.1999).

Anon. 1999b. Ireland - First national report on the implementation of the Convention on Biological Diversity. Government of Ireland. Dublin. 118 pp.

Barbero, M. Bonin, G., loisel, R. & Quézel, P. 1990. Changes and disturbances of forest ecosystems caused by human activities in the western part of Mediterranean basin. Vegetation 87: 151-173.

Barbosa, P.M., Pereira, J.M.C. and Grégoire, J.-M. 1998. Composting criteria for burned area assessment using multitemporal low resolution satellite data. Remote Sensing of Environment 65: 38-49.

Barbosa, P.M., Grégoire, J.-M. & Pereira, J.M.C. 1999. An Algorithm for Extracting Burned Areas from Time Series of AVHRR GAC Data Applied at a Continental Scale. Remote Sensing of Environment 69: 253-263.

Bartelink, H.H. & Olsthoorn, A.F.M. 1999. Introduction: mixed forest in western Europe. In: Olsthoorn, A.F.M., Bartelink, H.H., Gardiner, J.J., Pretzsch, H., Hekhuis, H.J. & Franc, A. (eds.): Management of mixed-species forest: silviculture and economics. IBN Scientific Contributions 15. IBN-DLO, Wageningen. P: 9-16.

Bengtsson, J., Nilsson, S.G., Franc, A. & Menozzi, P. 2000. Biodiversity, disturbances, ecosystem function and management of European forests. Forest Ecology and Management 132: 39-50.

Bibby, C.J., Collar, N.J., & Crosby, M.J., Heath, M. F., Imboden, C., Johnson, T. H., Long, A. J., Stattersfield, A. J., & Thirgood, S. 1992. Putting biodiversity on the map: Priority areas for global conservation. Cambridge, UK: International Council for Bird Preservation. 90 pp.

Bojinov, H., Stoichkova, P. & Sapunjieva, K. (eds.) 1998. National report for biological diversity conservation in Bulgaria. Bulgarian Ministry of Environment and Water. Sofia. 35 pp.

Brink, B. ten. 2000. Biodiversity indicators for the OECD environmental outlook and strategy. A feasibility study. RIVM Report 402001014, Global dynamics and sustainable development programme, Globo report series No. 25. 52 pp.

Bücking, W., Al., E., Falcone, P., Latham, J. & Sohlberg, S. 2000. Working Group I: Strict forest reserves in Europe and forests left to free development in other categories of protection. In: COST Action E4, Forest reserves research network. Office for Official Publications of the European Communities. Luxembourg. P: 39 – 134.

De Leo, G.A. & Levin, S. 1997. The multifaceted aspects of ecosystem integrity. Conservation Ecology [online] 1:3. http://www.consecol.org/vol1/iss1/art3.

De Long, D. 1996. Defining biodiversity. Wildlife Society Bulletin 24: 738-749.

Dudley, N. & Jeanrenaud, J.-P. 1998. Needs and prospects for international co-operation in assessing forest biodiversity: an overview from WWF. In Bachmann, P., Köhl, M. & Päivinen, R. (eds.). Assessment of biodiversity for improved forest planning. Proceedings of the conference on assessment of biodiversity for improved forest planning 7-9 October 1996, held in Monte Verità, Switzerland. European Forest Institute Proceedings No 18. Kluwer Academic Publishers. P: 31-41.

EEA 1995. Europe's Environment: The Dobris Assessment. European Environmental Agency. Copenhagen.

EEA 1999. Environment in the European Union at the turn of the century. Chapter 3.11. Changes and loss of biodiversity. European Environmental Agency. Copenhagen. P: 285-310.

EC 1995. Europe's Environment: Statistical compendium for the Dobris assessment. Office for Official Publications of the European Communities. 455 pp.

EC 1998. First report on the implementation of the convention on biological diversity by the European Community. European Commission. 82 pp.

EC and UN/ECE. 1999a. Forest Condition in Europe. 1999. Executive Report. Geneva and Brussels. 31 pp.

EC and UN/ECE. 1999b. De Fries, W., Reinds, G.J., Deelstra, H.D. Klap, J.M. & Vel, E.M. Intensive Monitoring of Forest Ecosystems in Europe. Technical Report 1999. Geneva and Brussels. 173 pp.

ENEA. 1998. Italian national report on the implementation status of the Convention on Biological diversity. Ministry of the Environment, Nature Conservation Service. Rome, 34 pp.

Ferris, R. & Humprey, J.W. 1999. A review of potential biodiversity indicators for application in British forests. Forestry 72: 313-328.

Fischer, A.G. 1961. Latitudinal variations in organic diversity. Am.Sci. 49: 50-79.

Franc, A. 1998. Some mathematical remarks on forest biodiversity. In: Bachmann, P., Köhl, M. & Päivinen, R. (eds.). Assessment of biodiversity for improved forest planning. Proceedings of the conference on assessment of biodiversity for improved forest planning 7-9 October 1996, held in Monte Verità, Switzerland. European Forest Institute Proceedings No 18. Kluwer Academic Publishers. P: 159-169.

Furness, R.W. & Greenwood, J.J.D. (eds). 1993. Birds as monitors of environmental change. Chapman & Hall, London. 360 pp.

Gabbay, S. (ed.) 1997. Conservation and sustainable use of biological diversity in Israel. Report of the state of Israel on the implementation of Article 6 of the Convention on Biological Diversity. State of Israel, Ministry of Environment. 96 pp.

Gadow, K.v. & Puumalainen, J. 1998. Neue Herausforderungen für die Waldökosystemplanung. (Summary: New challenges for the forest ecosystem management). Allgemeine Forstzeitschrift/Der Wald 20: 1248 – 1250.

Grabherr, G. 1997. Naturschutzfachliche Bewertung der Natürlichkeit österreichischer Wälder. ÖFZ 1: 11-12.

Grumbine, E.R. 1994. What is ecosystem management? Conservation Biology 8: 27-38.

Haila, Y. 1983. Land birds on northern islands: a sampling metaphor for insular colonization. Oikos 41: 334-351.

Hannah, L., Carr, J.L. & Lankerani, A. 1995. Human disturbance and natural habitat: a biome level analysis of a global data set. Biodiversity Conservation 4: 128-155.

Hanski, I. 2000: Extinction debt and species credit in boreal forests: modelling the consequences of different approaches to biodiversity conservation. — Ann. Zool. Fennici 37: 271–280.

Hanski, I. 1999. Metapopulation Ecology. Oxford series in Ecology and Evolution, Oxford University Press Inc. New York. 313 pp.

Hellawell, J.W. 1991. Development of a rationale for monitoring. In: Goldsmith, F.B. (ed.). Monitoring for conservation and ecology. Chapmann and Hall, London. P: 1-14.

Hubbel, S.P. & Foster, R.B. 1983. Diversity of canopy trees in a neotropical forest and implications for conservation. In Sutton, S.L., Whitmore, T.C. & Chadwick, A.C. (eds.): Tropical rain forest: Ecology and management. Blackwell Scientific, Oxford. P: 25-41.

Jaakkola, S. 1998. UNEP's global biodiversity assessment. In Bachmann, P., Köhl, M. & Päivinen, R. (eds.) 1998. Assessment of biodiversity for improved forest planning. Proceedings of the Conference on assessment of biodiversity for improved forest planning 7-9 October 1996, held in Monte Verità, Switzerland. European Forest Institute Proceedings No 18. Kluwer Academic Publishers. P: 43-50.

Jeffers, J.N.R. 1996. Measurement and characterisation of biodiversity in forest ecosystems – new methods and models. In: Bachmann, P., Kuusela, K. & Uuttera, J. (eds.). Assessment of biodiversity for improved forest management. EFI Proceedings No 6: 59-67.

Kabucis, I., Opermanis, O., Lodzina, I., Tesnova, I. & Spungis, V. 1998. National Report on biological diversity: Latvia. Ministry of Environmental Protection and Regional Development & UNEP. Riga. 49 pp.

Kapos, V. & Iremonger, S.F. 1998. Achieving global and regional perspectives on forest biodiversity and concervation. In Bachmann, P., Köhl, M. & Päivinen, R. (eds.) Assessment of biodiversity for improved forest planning. Proceedings of the conference on assessment of biodiversity for improved forest planning 7-9 October 1996, held in Monte Verità, Switzerland. European Forest Institute Proceedings No 18. Kluwer Academic Publishers. P: 3-13.

Klinda, J. 1998. Environment of the Slovak Republic. Paper prepared for the 4th meeting of the Conference of the Parties to the Convention on Biological Diversity. Ministry of the Environment of the Slovak Republic and Slovak Environmental Agency. Bratislava. 142 pp.

Kouki, J. (ed.) 1994. Biodiversity in the Fennoscandian boreal forests: natural variation and its management. Ann. Zool. Fennici 31. 217 pp.

Kull, T. (ed.). 1999. Estonian Biodiversity strategy and action plan. Estonian Ministry of Environment, UNEP and Environmental Protection Institute of Estonian Agricultural University. Tallinn – Tartu. 165 pp.

Kuusela, K. 1994. Forest resources in Europe, 1950-1990. European Forest Institute, Research Report 1. Cambridge University Press. 154 pp.

Larsson, T.-B., Svensson, L., Angelstam, P., Balent, G., Barbati, A., Bijlsma, R.-J., Boncina, A., Bradshaw, R., Bücking, W., Ciancio, O., Corona, P., Diaci, J., Dias, S., Ellenberg, H., Fernendes, F.M., Fernández-Gonzalez, F., Ferris, R., Frank, G., Møller, P.F., Giller, P.S., Gustafsson, L., Halbritter, K., Hall, S., Hansson, L., Innes, J., Jactel, H., Keannel Dobbertin, M., Klein, M., Marchetti, M., Mohren, F., Niemelä, P. O'Halloran, J., Rametsteiner, E., Rego, F., Scheidegger, C., Scotti, R., Sjöberg, K., Spanos, I., Spanos, K., Standovár, T., Tømmerås, Å., Trakolis, D., Uuttera, J., Walsh, P.M., Vandekerkhove, K., Watt, A.D. & VenDenMeersschaut, D. 2001. Biodiversity evaluation tools for European forests. A report from the FAIR project "Indicators for monitoring and evaluation of forest biodiversity in Europe" CT97-3575 within the EU Commission RTD Programme. 154 pp. & Appendices. Ecological Bulletins 50. In press.

Legakis, A. & Spyropoulou, S. (eds.). 1998. First national report on the convention on biological diversity. Greece. Ministry of environment, physical planning and public works, General directorate for environment, Natural environment management section. Athens. 50 pp.

Leikola, M. 1999. Definition and classification of mixed forests, with a special emphasis on boreal forests. In: Olsthoorn, A.F.M., Bartelink, H.H., Gardiner, J.J., Pretzsch, H., Hekhuis, H.J. & Franc, A. (eds.): Management of mixed-species forest: silviculture and economics. IBN Scientific Contributions 15. IBN-DLO, Wageningen. P: 20-28.

MacArthur, R.H. & MacArthur, J.W. 1961. On bird species diversity. Ecology 42: 594-598.

MacArthur, R.H. & Wilson, E.O. 1967. The theory of island biogeography. Monographs in Population Biology 1. Princeton University Press. Princeton, NJ. 199 pp.

Marhold, K. & Hindak, F. (eds.) 1998. Checklist of non-vascular and vascular plants of Slovakia. Veda, Publishing house of SAV. Bratislava. 687 pp.

Matevski, V., Bukleska-Ralevska, A & Nastov, A. (eds.) 2000. Republic of Macedonia, Ministry of Environment. National report on biodiversity. Skopje, Macedonia.

McCormick, N. & Folving, S. 1998. Monitoring European forest biodiversity at regional scales using satellite remote sensing. In Bachmann, P., Köhl, M. & Päivinen, R. (eds.) 1998. Assessment of biodiversity for improved forest planning. Proceedings of the Conference on assessment of biodiversity for improved forest planning 7-9 October 1996, held in Monte Verità, Switzerland. European Forest Institute Proceedings No 18. Kluwer Academic Publishers. P: 283-289.

MCPFE. 2000. General declarations and resolutions adopted at the Ministerial Conferences on the protection of forests in Europe. Strasbourg 1990 – Helsinki 1993 – Lisbon 1998. Ministerial Conference on the Protection of Forests in Europe. Liaison Unit Vienna. 88 pp.

Mikusiński, G. & Angelstam, P. 1998. Economic geography, forest distribution, and woodpecker diversity in Central Europe. Conservation Biology 12: 200-208.

Møller, P.F. 1997. Biodiversity in Danish natural forests. A comparison between unmanaged and managed forests in East Denmark. (in Danish with English summary and subtext). Danmarks Geologiske Undersøgelse Rapport 1997/41. 209 pp.

Murdoch, W.W., Evans, F.C. & Peterson, C.H. 1972. Diversity and pattern in plants and insects. Ecology 53: 819-828.

Myers, N., Mittelmeier, R.A., Mittelmeier, C.G., da Fonseca, G.A.B. & Kent, J. 2000. Biodiversity hotspots for conservation priorities. Nature 403: 853-858.

Niemelä, J. 2000: Biodiversity monitoring for decision-making. — Ann. Zool. Fennici 37: 307–317.

Noss, R.F. 1990. Indicators for monitoring biodiversity: A hierarchical approach. Essay. Conservation Biology 4: 355-364.

Noss, R.F. 1999. Assessing and monitoring forest biodiversity. A suggested framework and indicators. Forest Ecology and Management 115: 135-146.

Nowicki, P., Saether, B.-E. & Solhaug, T. 1998. Threats to biodiversity. In: Catizzone, M., Larsson, T.B. & Svensson, L. (eds.) "Understanding biodiversity, a research agenda prepared by the European Working Group on Research and Biodiversity. European Commission, Ecosystems Res. Rep. 25. P: 25-31.

OECD 1993. OECD Environmental Performance Reviews – Norway. OECD Publications, Paris. 162 pp.

OECD 1994. OECD Environmental Performance Reviews – Italy . OECD Publications, Paris. 207 pp.

Päivinen, R, Lehikoinen, M. Schuck, A., Häme, T., Väätäinen, S., Sedano, F. & Sirro, L. 2000. Combining geographically referenced Earth observation data and forest statistics for deriving a forest map for Europe.(Final Report of the EC Contract No. 15237-1999-08F1EDISPFI). European Forest Institute in association with VTT Automation, University of Joensuu - Faculty of Forestry and Stora Enso Forest Consulting. Internal documentation. 118 pp.

Päivinen, R. & Köhl, M. (eds.) 1997. Study on forestry information and communication system – Reports on forest inventory and survey systems. Vol. 1&2. Office for Official Publications of the European Communities, Luxembourg. 1328 pp.

Palmer, M.W. & White, P.S. 1994. Scale dependence and the species-area relationship. American Naturalist 144: 717-740.

Parviainen, J., Little, D., Doyle, M., O'Sullivan, A., Kettunen, M. & Korhonen, M. (eds.) 1999. Research in forest reserves and natural forests in European countries. Country reports for the COST Action E4: Forest Reserves Research Network. EFI Proceedings No. 16. European Forest Institute, Joensuu. 299 pp.

Parviainen, J., Kassioumis, K., Bücking, W., Hochbichler, E., Päivinen, R. & Little, D. 2000. Final report summary: Mission, goals, outputs, linkages, recommendations and partners. In: COST Action E4, Forest reserves research network. Office for Official Publications of the European Communities. Luxembourg. P: 9–38.

Peet, R.K. 1978. Forest vegetation of the Colorade front range: Patterns of species diversity. Vegetatio 37: 65-78.

Peterson, K. 1994. Nature conservation in Estonia – general data and protected areas. Estonian Ministry of Environment. Huma Publishers, Tallinn. 48 pp.

Peterson, K., Marald, T., Kuldna. P. (eds.) 1998. First national report to the convention on biological diversity in Estonia. Estonian Ministry on Environment. 29 pp.

Plesnik, J. & Roudna, M. (eds.) 1999. National Biodiversity Conservation Strategy and Action Plan in the Czech Republic - Status of Biological Resources and Implementation of the Convention on Biological Diversity in the Czech Republic - First Report. Prague.

Pretzsch, H. 1997. Analysis and modelling of spatial stand structures. Methodological considerations based on mixed beech-larch stands in Lower Saxony. Forest Ecology and Management 97: 237-253.

Pretzsch, H. 1999a. Structural diversity as a result of silvicultural operations. In: Olsthoorn, A.F.M., Bartelink, H.H., Gardiner, J.J., Pretzsch, H., Hekhuis, H.J. & Franc, A. (eds.): Management of mixed-species forest: silviculture and economics. IBN Scientific Contributions 15. IBN-DLO, Wageningen. P: 157-174.

Pretzsch, H. 1999b. Waldwachstum im Wandel. Summary: Changes in forest growth. Forstw. Cbl. 118: 1-23.

Quézel, P., Médail, F., Loisel, R. & Barbero, M. 1999. Biodiversity and conservation of forest species in the Mediterranean basin. Unasylva Vol 50. No. 197: 21-28.

SAEFL. 1998. National report of Switzerland for the Convention on Biological Diversity. Swiss Agency for the Environment, Forests and Landscape (SAEFL). Hintermann & Weber SA. Montreux/Bern. 64 pp.

Solberg, B. 1998. Biodiversity protection and forest management – some economic and policy aspects. In Bachmann, P., Köhl, M. & Päivinen, R. (eds.) Assessment of biodiversity for improved forest planning. Proceedings of the conference on assessment of biodiversity for improved forest planning 7-9 October 1996, held in Monte Verità, Switzerland. European Forest Institute Proceedings No 18. Kluwer Academic Publishers. P: 53-62.

Spiecker, H., Mielikäinen, K., Köhl, M. & Skovsgaard, J.P. (eds.) 1996. Growth trends in European forests. EFI Research Report 5. Springer-Verlag. Heidelberg – Berlin. 372 pp.

Stevanovic, V. & Vasic, V. (eds.) 1995. Biodiversity of Yugoslavia with an overview of internationally important species (in Serbian). Faculty of Biology and Ecolibri. Belgrade. 554 pp.

Swaay, C.A.M. Van & Warren, M.S. 1999. Red data book of European butterflies. Nature and Environment, No. 99, Council of Europe Publishing, Strasbourg.

Third Ministerial Conference on the Protection of Forests in Europe. 1998. Follow-up reports, Volume II: Sustainable forest management in Europe. Special Report on the follow-up on the implementation of resolutions H1 and H2 of the Helsinki Ministerial Conference. Published by the Ministry of Agriculture, Rural Development and Fisheries of Portugal. Lisbon. 274 pp.

Thomas, J.A. (1995). The conservation of declining butterfly populations in Britain and Europe: Priorities, problems and successes. Biological Journal of the Linnean Society 56: 55-72.

Thorne, C. & Isermeyer, F. 1997. Biologische Vielfalt und ökonomische Rahmenbedingungen. Summary: Biological diversity and economic basic conditions. Schriftenreihe des BML "Angewandte Wissenschaft", Band 465 "Biologische Vielfalt in Ökosystemen": 210 - 225.

Trpin, D. & Vres, B. 1995. Register of the Flora of Slovenia. Ferns and Vascular plants. SAZU, Ljubljana. 143 pp.

UN 2000. Forest resources of Europe, CIS, North America, Australia, Japan and New Zealand (industrialized temperate/boreal countries). UN/ECE/FAO Contribution to the Global Forest Resources Assessment 2000. Main Report. Geneva Timber and Forest Study Papers, No. 17. United Nations Publication, New York and Geneva. 445 pp.

UNEP 1997. Recommendations for a core set of indicators of biological diversity. Convention of Biological diversity UNEP/CBD/SBSTTA/3/9. Montreal.

UNEP 1999. Global Environmental Outlook 2000. Produced by the United Nations Environment Programme. Earthscan Publications Ltd. London. Available also: http://www.grida.no/geo2000/english/index.htm.

Vanclay, J.K. 1998. Towards more rigorous assessment of biodiversity. In: Bachmann, P., Köhl, M. & Päivinen, R. (eds.). Assessment of biodiversity for improved forest planning. Proceedings of the conference on assessment of biodiversity for improved forest planning 7-9 October 1996, held in Monte Verità, Switzerland. European Forest Institute Proceedings No 18. Kluwer Academic Publishers. P: 211-232.

WCMC (comp.) Groombridge, B. (ed.) 1994. Biodiversity data sourcebook. WCMC Biodiversity series No 1. World Conservation Press, Cambridge, UK. 155 pp.

Weider, L. J. & Hobæk, A. 2000. Phylogeography and arctic biodiversity: a review. — Ann. Zool. Fennici 37: 217–231.

9. Some Selected Web-addresses Related to Forest Biodiversity

The wealth of biodiversity information in the Internet is enormous. Thus, the following links in no way make a comprehensive overview of possible information sources and are provided just to serve as a starting point for those interested. Generally, information of differing quality is available in Internet – also on forest biodiversity.

UNECE TIMBER COMMITTEE Temperate and Boreal Forest Resources (The full TBFRA-2000 report)

http://www.unece.org/trade/timber/fra/welcome.htm

Joint Research Centre of the European Commission

http://www.jrc.org/

Pan-European Process, i.e. Ministerial Conference on the Protection of Forests in Europe

http://www.minconf-forests.net/

Bear-project (Larsson et al. 2001)

http://www.algonet.se/~bear/

Convention on biological diversity and its clearing-house mechanism

http://www.biodiv.org/chm/

European Community Biodiversity Strategy

http://europa.eu.int/comm/development/sector/environment/env_theme/biodiversity/ec_policy/txt01.pdf

European Community Biodiversity Action Plans in the areas of Conservation of Natural Resources, Agriculture, Fisheries, and Development and Economic Co-operation

http://europa.eu.int/comm/environment/biodiversity/index_en.htm

European Community Biodiversity Clearing House managed by the European Environmental Agency

http://biodiversity-chm.eea.eu.int/

CORDIS – European Community Research and Development Information System

http://www.cordis.lu/

Pan-European Biological and Landscape Diversity Strategy

http://www.ecnc.nl/doc/europe/legislat/strafull.html

UNEP Global Environmental Outlook (GEO) - A new State of the Environment Report

http://www.unep.org/unep/eia/geo/intro.htm

FAO forestry web-sites

http://www.fao.org/forestry/fo/fra/index.jsp

World Conservation Monitoring Centre WCMC

http://www.unep-wcmc.org.

World Resources Institute WRI

http://www.wri.org/wri/biodiv/

COST ACTION E4 "Forest Reserves Research Network"

http://www.efi.fi/Database_Gateway/FRRN/news.html

European Forest Institute

http://www.efi.fi/

COST – European Cooperation in the Field of Scientific and Technical Research

http://www.belspo.be/cost

European Working Group on Research and Biodiversity (EWGRB)

http://europa.eu.int/comm/research/envir/ewgrb.html

IUCN

http://www.iucn.org/

United Nations Environment Programme

http://www.unep.org/

The Bern Convention

http://www.nature.coe.int/english/cadres/berne.htm

European Centre for Nature Conservation

http://www.ecnc.nl/

Biodiversity Assessment Tools - Identifying Indicators to assess the impact of European Policies on Biodiversity (Electronic Conference and Project)

http://www.gencat.es/mediamb/bioassess/

Biodiversity in Central and Eastern Europe, UNEP & GRIDArendal

http://www.grida.no/enrin/biodiv/index_en.htm

SOME FACTS ABOUT THE JOINT RESEARCH CENTRE

The Joint Research Centre is the European Union's scientific and technical research laboratory and an integral part of the European Commission. JRC provides the scientific advice and technical know-how to support EU policies. "The mission of the JRC is to provide customer-driven scientific and technical support for the conception, development, implementation and monitoring of EU policies. As a service of the European Commission, the JRC functions as a reference centre of science and technology for the Union. Close to the policy-making process, it serves the common interest of the Member States, while being independent of special interests, whether private or national."

The structure of the JRC is based on eight specialised institutes, with research co-ordinated by the JRC Programmes Directorate. The JRC employs about 2200 staff (end of 2000) and uses a budget of over 300 million Euro per year stemming from the European Commission's research budget and from competitive income. The institutes are located on five separate sites in Belgium, Germany, Italy, the Netherlands and Spain.

Over 95% of the JRC activities are currently focused on four key areas, namely on safety of food and chemical products, environment, dependability of information systems and services, and nuclear safety and safeguards. In these areas, the JRC assists to draw up regulations and develop test methods as well as to ensure greater safety and security for European citizens and greater competitiveness for European industry. The guideline is that of 'adding value' where appropriate, rather than competing directly with establishments in the EU Member States.

More information about the work of the JRC and Environment and GEOinformation Unit may be obtained by contacting:

JRC - European Commission

Information and Public Relations

I – 21020 Ispra (Va), Italy

Tel: +39 0332 789893

Fax: +39 0332 782435

E-mail: Ulla.Engelmann@cec.eu.int

WEB site address: http://www.jrc.cec.eu.int

SOME FACTS ABOUT THE TIMBER COMMITTEE

The Timber Committee is a principal subsidiary body of the ECE (UN Economic Commission for Europe) based in Geneva. It constitutes a forum for cooperation and consultation between member countries on forestry, forest industry and forest product matters. All countries of Europe; the former USSR; United States of America, Canada and Israel are members of the ECE and participate in its work.

The ECE Timber Committee shall, within the context of sustainable development, provide member countries with the information and services needed for policy- and decision-making regarding their forest and forest industry sector ("the sector"), including the trade and use of forest products and, when appropriate, formulate recommendations addressed to member Governments and interested organizations. To this end, it shall:

1. With the active participation of member countries, undertake short-, medium- and long-term analyses of developments in, and having an impact on, the sector, including those offering possibilities for the facilitation of international trade and for enhancing the protection of the environment;
2. In support of these analyses, collect, store and disseminate statistics relating to the sector, and carry out activities to improve their quality and comparability;
3. Provide the framework for cooperation e.g. by organizing seminars, workshops and ad hoc meetings and setting up time-limited ad hoc groups, for the exchange of economic, environmental and technical information between governments and other institutions of member countries that is needed for the development and implementation of policies leading to the sustainable development of the sector and to the protection of the environment in their respective countries;
4. Carry out tasks identified by the UNECE or the Timber Committee as being of priority, including the facilitation of subregional cooperation and activities in support of the economies in transition of central and eastern Europe and of the countries of the region that are developing from an economic point of view;
5. It should also keep under review its structure and priorities and cooperate with other international and intergovernmental organizations active in the sector, and in particular with the FAO (Food and Agriculture Organization) and its European Forestry Commission and with the ILO (International Labour Organisation), in order to ensure complementarity and to avoid duplication, thereby optimizing the use of resources.

More information about the Committee's work may be obtained by contacting:

Timber Section
UNECE Trade Division
Palais des Nations
CH - 1211 GENEVA 10, Switzerland
Fax: +41 22 917 0041
E-mail: info.timber@unece.org

WEB site address: http://www.unece.org/trade/timber

UNECE/FAO Publications

Timber Bulletin Volume LIII (2000) ECE/TIM/BULL/53/...

Timber Bulletin Volume LIV (2001) ECE/TIM/BULL/54/...

1. Forest Products Prices
2. Forest Products Statistics (database [chronological series, since 1964] also available on diskettes)
3. Forest Products Annual Market Review
4. Forest Fire Statistics
5. Forest Products Trade Flow Data
6. Forest Products Markets in *(current year)* and Prospects for *(forthcoming year)*

Geneva Timber and Forest Study Papers

Forest policies and institutions in Europe, 1998-2000	ECE/TIM/SP/19
Forest and Forest Products Country Profile: Russian Federation	ECE/TIM/SP/18
Forest Resources of Europe, CIS, North America, Australia, Japan and New Zealand	ECE/TIM/SP/17
State of European forests and forestry, 1999	ECE/TIM/SP/16
Non-wood goods and services of the forest	ECE/TIM/SP/15
European Timber Trends and Prospects: into the 21st century (ETTS V)	ECE/TIM/SP/11

The above series of sales publications and subscriptions are available through United Nations Publications Offices as follows:

Orders from Africa, Europe and the Middle East should be sent to:
Sales and Marketing Section, Room C-113
United Nations
Palais des Nations
CH - 1211 Geneva 10, Switzerland
Fax: + 41 22 917 0027
E-mail: unpubli@unog.ch
Web site: http://www.un.org/Pubs/sales.htm

Orders from North America, Latin America and the Caribbean, Asia and the Pacific should be sent to:
Sales and Marketing Section, Room DC2-853
United Nations
2 United Nations Plaza
New York, N.Y. 10017, United States of America
Fax: + 1 212 963 3489
E-mail: publications@un.org

* * * * *

Geneva Timber and Forest Discussion Papers *(original language only)*

Markets for Secondary Processed Wood Products, 1990-2000	ECE/TIM/DP/21
Forest Certification update for the ECE Region, summer 2000	ECE/TIM/DP/20
Trade and environment issues in the forest and forest products sector	ECE/TIM/DP/19
Multiple use forestry	ECE/TIM/DP/18
Forest Certification update for the ECE Region, summer 1999	ECE/TIM/DP/17
A summary of "The competitive climate for wood products and paper packaging: the factors causing substitution with emphasis on environmental promotions"	ECE/TIM/DP/16
Recycling, Energy and Market Interactions	ECE/TIM/DP/15
The Status of forest certification in the ECE region	ECE/TIM/DP/14
The role of women on forest properties in Haute-Savoie: Initial researches	ECE/TIM/DP/13
Interim Report on the Implementation of Resolution H3 of the Helsinki Ministerial Conference on the protection of forests in Europe (Results of the second enquiry)	ECE/TIM/DP/12
Manual on acute forest damage	ECE/TIM/DP/7

International Forest Fire News *(two issues per year)*

Timber and Forest Information Series

Timber Committee Yearbook 2001 ECE/TIM/INF/8

The above series of publications may be requested free of charge through:

Timber Section
UNECE Trade Division
United Nations
Palais des Nations
CH - 1211 Geneva 10, Switzerland
Fax: + 41 22 917 0041
E-mail: info.timber@unece.org

Downloads are available at http://www.unece.org/trade/timber